Information Circular 9520

One Hundred Years of Federal Mining Safety and Health Research

By John A. Breslin, Ph.D.

DEPARTMENT OF HEALTH AND HUMAN SERVICES
Centers for Disease Control and Prevention
National Institute for Occupational Safety and Health
Pittsburgh Research Laboratory
Pittsburgh, PA

February 2010

This document is in the public domain and may be freely copied or reprinted.

Disclaimer

Mention of any company or product does not constitute endorsement by the National Institute for Occupational Safety and Health (NIOSH). In addition, citations to Web sites external to NIOSH do not constitute NIOSH endorsement of the sponsoring organizations or their programs or products. Furthermore, NIOSH is not responsible for the content of these Web sites.

Ordering Information

To receive documents or other information about occupational safety and health topics, contact NIOSH at

>Telephone: **1–800–CDC–INFO** (1–800–232–4636)
>TTY: 1–888–232–6348
>e-mail: cdcinfo@cdc.gov
>
>or visit the NIOSH Web site at **www.cdc.gov/niosh**.

For a monthly update on news at NIOSH, subscribe to NIOSH *eNews* by visiting **www.cdc.gov/niosh/eNews**.

DHHS (NIOSH) Publication No. 2010-128

February 2010

SAFER • HEALTHIER • PEOPLE™

CONTENTS

Page

1. Introduction — 1

2. The Beginning: 1910-1925 — 1
   - 2.1 Early Legislation — 1
   - 2.2 Early Mine Disasters — 2
   - 2.3 The Origins of the USBM and the Organic Act — 7
     - 2.3.1 Initial Safety Research — 9
   - 2.4 The USBM and Miner Health — 10
   - 2.5 The USBM Opens an Experimental Mine near Pittsburgh — 12
   - 2.6 Expansion of the USBM Mission — 15
   - 2.7 Expansion of the USBM in Pittsburgh — 17
   - 2.8 Early Statistics on Mining Fatalities and Production — 18
   - 2.9 The USBM and WW I — 19

3. Hard Times and Maturation: 1925-1969 — 21
   - 3.1 USBM Transferred to the Department of Commerce — 21
     - 3.1.1 Explosives Research — 21
     - 3.1.2 Rock Dusting Research — 22
     - 3.1.3 Safety Inspection — 23
   - 3.2 USBM Transferred Back to the Department of the Interior — 23
   - 3.3 Impact of the Great Depression on the USBM — 24
   - 3.4 The USBM and WW II — 24
   - 3.5 Significant USBM Contributions Outside of Mining — 25
     - 3.5.1 Explosives and Flammability Research — 25
     - 3.5.2 Metals Research — 26
   - 3.6 Mechanization of Coal Mining over the Course of the Century — 27
     - 3.6.1 Roof Control — 28
   - 3.7 The Federal Coal Mine Safety Act of 1952 — 30
   - 3.8 Permissible Explosives Research — 31

4. Public Health Service Mining Studies — 32
   - 4.1 Joint Public Health Service and USBM Studies — 32
     - 4.1.1 Silicosis — 32
     - 4.1.2 The Uranium Mining Study — 33
     - 4.1.3 Coal Miner Health Studies — 33
   - 4.2 The USBM in the 1960s — 34
   - 4.3 USBM Research Laboratories in the 1960s — 36
     - 4.3.1 Denver Mining Research Center, Denver, CO — 36
     - 4.3.2 Applied Physics Research Laboratory (Mining), College Park, MD — 36
     - 4.3.3 Minneapolis Mining Research Center, Minneapolis, MN — 36
     - 4.3.4 Reno Mining Research Laboratory, Reno, NV — 36
     - 4.3.5 Explosives Research Center, Mining Research Center, Health and Safety Research Testing Center, Pittsburgh, PA — 37
     - 4.3.6 Bruceton Laboratories, Bruceton, PA — 37

4.3.7 Spokane Mining Research Laboratory, Spokane, WA	37
4.4 Legislation and USBM Accomplishments of the 1960s	37
5. Expansion and Change: 1969-1996	**39**
5.1 The Federal Coal Mine Health and Safety Act of 1969	39
5.2 Establishment of NIOSH	41
5.3 Dismemberment of the USBM	43
5.3.1 Transfer of USBM Research Labs to ERDA	46
5.3.2 Coal Advanced Mining Technology Program	46
5.4 Federal Mine Safety and Health Act of 1977	47
5.5 USMB Research Centers after 1979 Reorganization	49
5.5.1 Pittsburgh Research Center	49
5.5.2 Twin Cities Research Center	50
5.5.3 Spokane Research Center	51
5.5.4 Denver Research Center	52
5.5.5 Mineral Institutes	52
5.6 USBM Accomplishments in Mining Research	53
5.6.1 Respirable Dust	53
5.6.2 Ground Control	54
5.6.3 Disaster Prevention	56
5.6.4 Illumination	57
5.7 R&D 100 awards and the USBM	58
5.8 Decline and Dissolution of the USBM	58
6. Consolidation and New Beginning: 1996-2010	**61**
6.1 Mining Research under NIOSH	61
6.2 NIOSH Mining-Related Research Accomplishments	63
6.2.1 Permissible Exposure Limit/Recommended Exposure Limit for Coal Mine Dust	63
6.2.2 Coal Workers' Pneumoconiosis (CWP)	63
6.2.3 Personal Dust Monitor (PDM)	64
6.2.4 Improved Regulations for Dust Control Associated with Highwall Drilling at Surface Coal Mines	65
6.2.5 Diesel Particulate Regulations	66
6.2.6 Epidemiologic Studies of the Vermiculite Mine at Libby, MT	67
6.2.7 Reducing Noise on Mining Machines	68
6.2.8 Ground Control	70
6.2.9 Coal Mine Roof Rating System	70
6.2.10 Guidelines for Coal Pillar Recovery	71
6.2.11 Support Technology Optimization Program (STOP) Design Software	71
6.3 Advancements in Mine Fires and Explosions Research	72
6.3.1 Coal Dust Explosibility Meter	72
6.3.2 Improving Underground Coal Mine Sealing Strategies	73
6.3.3 Refuge Chambers	73

    6.4 Advancements in Injury Prevention and Research   74
        6.4.1 Hazardous Area Signaling and Ranging Device (HASARD)—
            Proximity Warning System   74
        6.4.2 Improved Training Materials and Methods   75
    6.5 Current NIOSH Mining Program   75
        6.5.1 Partnerships   76
        6.5.2 NAS Evaluation of NIOSH Mining Program   76
    6.6 Recent External Factors Affecting Mining Research   77

7. Conclusion: Lessons Learned   78

8. References   80

ILLUSTRATIONS

Figure 1. Young boys working in tipple, a type of rudimentary coal screening plant.   2
Figure 2. Child miner wearing cloth cap with open flame oil cap lamp, circa 1900.   3
Figure 3. Headline from the 1907 Monongah coal mine disaster.   4
Figure 4. Peter Urban, Monongah Mine disaster survivor.   4
Figure 5. Outdoor morgue following Monongah explosion.   5
Figure 6. Plaque commemorating the Darr Mine disaster.   6
Figure 7. Rescuers (nicknamed "helmet men" because of the devices they wore) equipped with newly developed air supply apparatus at the Cherry Mine disaster.   7
Figure 8. Official seal of the United States Bureau of Mines, established in 1910.   8
Figure 9. Photo of Dr. Joseph A. Holmes.   8
Figure 10. Dr. Joseph A. Holmes and miners being trained in front of mine rescue car, 1913.   9
Figure 11. Miner preparing explosive shot in coal mine.   10
Figure 12. Early airborne dust sampler developed by the USBM, circa 1915.   11
Figure 13. Early view of the USBM's Bruceton, PA, facility.   12
Figure 14. President William H. Taft and Dr. Joseph A. Holmes at national mine safety demonstration, Pittsburgh, 1911.   13
Figure 15. Public demonstration of coal dust explosion at Bruceton experimental mine.   13
Figure 16. U.S. Bureau of Mines exhibit at San Francisco Exposition, CA, 1915.   14
Figure 17. Mine rescue exhibition at Forbes Field, Pittsburgh, PA, circa 1918.   14
Figure 18. USBM exhibit at Century of Progress exhibition in Chicago, IL, 1933.   14
Figure 19. Collapsible cage used for mine rescue.   17
Figure 20. The USBM's Forbes Avenue facility under construction, Pittsburgh, PA, circa 1915.   18
Figure 21. The USBM's completed Forbes Avenue building, Pittsburgh, PA, 1917.   18
Figure 22. Miners wearing the Gibbs mine rescue breathing apparatus, 1918.   20
Figure 23. Ford Model T in the Bruceton Experimental Mine, 1921.   21
Figure 24. Explosion test at the Bruceton Experimental Mine, 1927.   22
Figure 25. Undercutting coal with a hand pick in the early 20th century.   27

Figure 26. Joy loading machine in 1925.   28
Figure 27. Coal mining production by mining method in the 20th century.   28
Figure 28. Setting timber posts in a coal mine for roof control.   29
Figure 29. The impact of roof bolts in the Isabella Mine, Isabella, PA, in 1949.   29
Figure 30. The results of conventional timbering at the Isabella Mine, Isabella, PA, in 1949.   30
Figure 31. John L. Lewis, President of the United Mine Workers of America, at West Frankfort disaster site, 1951.   30
Figure 32. Smoke from the Farmington Mine disaster, Farmington, WV, 1968.   39
Figure 33. Aerial view of the Bruceton facility in PA, 1988.   49
Figure 34. Main portal for the Lake Lynn Laboratory near Fairchance, PA.   50
Figure 35. Spokane Research Center, Spokane, WA, circa 1985.   51
Figure 36. Air currents when using the shearer-clearer system.   54
Figure 37. Testing of the mine roof simulator.   56
Figure 38. Percentage of miners with CWP by tenure in mining, CWXSP, 1970-2006.   63
Figure 39. Miner wearing personal dust monitor (PDM).   65
Figure 40. Collapsible Drill Steel Enclosure at start of drilling.   69
Figure 41. Coal Dust Explosibility Meter.   72
Figure 42. NIOSH Exhibit at the Society for Mining, Metallurgy, and Exploration Annual Meeting, Denver, 2009.   78

APPENDICES

Appendix A: Joseph A. Holmes Biography   85
Appendix B: Directors of USBM and NIOSH   87

TABLES

Table 1. U.S. Bureau of Mines appropriations and expenditures, fiscal years ending June 30, 1911-35   88
Table 2. U.S. Coal Fatalities, 1900 through 2008.   89
Table 3. Metal/Nonmetal Fatalities, 1900 through 2008.   90

ACRONYMS AND ABBREVIATIONS USED IN THIS REPORT

AEC	Atomic Energy Commission
AHSM	Analysis of Horizontal Stress in Mines
ALFORD	Appalachian Laboratory for Occupational Respiratory Diseases
ALOSH	Appalachian Laboratory for Occupational Safety and Health
ALPS	Analysis of Longwall Pillar Stability
ARMPS	Analysis of Retreat Mining Pillar Stability
ATSDR	Agency for Toxic Substances and Disease Registry
BETC	Bartlesville Energy Technology Center

BOSH	Bureau of Occupational Safety and Health
CDEM	Coal Dust Explosibility Meter
CDSE	Collapsible Drill Steel Enclosure
CFR	Code of Federal Regulations
CMRR	Coal Mine Roof Rating
CPDM	continuous personal dust monitor
CRS	Congressional Research Service
CWP	coal workers' pneumoconiosis
CWXSP	Coal Workers' X-ray Surveillance Program
DOE	Department of Energy
DOI	Department of the Interior
DPM	diesel particulate matter
DRDS	Division of Respiratory Disease Studies
DSR	Division of Safety Research
EC	elemental carbon
ERDA	Energy Research and Development Administration
GMTC	Generic Mineral Technology Center
HASARD	Hazardous Area Signaling and Ranging Device
HEW	Department of Health, Education and Welfare
HHS	Department of Health and Human Services
JAHSA	Joseph A. Holmes Safety Association
MESA	Mining Enforcement and Safety Administration
METC	Morgantown Energy Technology Center
METF	Mining Equipment Test Facility
MFIRE	mine fire simulation computer program
MHRAC	Mine Health Research Advisory Committee
MPPCF	million particles per cubic foot of air
MRS	mobile roof supports
MSHA	Mine Safety and Health Administration
MSHRAC	Mine Safety and Health Research Advisory Committee
NASA	National Aeronautics and Space Administration
NIOSH	National Institute for Occupational Safety and Health
NPPTL	National Personal Protective Technology Laboratory
NRC	National Research Council
OC	organic carbon
OMB	Office of Management and Budget
OSHA	Occupational Safety and Health Administration
OSM	Office of Surface Mining, Reclamation, and Enforcement
PDM	Personal Dust Monitor
PEL	permissible exposure limit
PETC	Pittsburgh Energy Technology Center
PHS	Public Health Service
PIB	Program Information Bulletin
PMF	progressive massive fibrosis
PRL	Pittsburgh Research Laboratory
R&D	research and development

r2p	research to practice
REL	recommended exposure limit
SCSR	self-contained self-rescuer
SMCRA	Surface Mining Control and Reclamation Act
SRL	Spokane Research Laboratory
STOP	Support Technology Optimization Program
TC	total carbon
U.S.	United States
USBM	U.S. Bureau of Mines
USGS	U.S. Geological Survey
WL	Working Level
WW I	World War I
WW II	World War II

SHORTENED FORMS USED IN THIS REPORT

1969 Coal Act	Federal Coal Mine Health and Safety Act of 1969
1977 Mine Act	Federal Mine Safety and Health Act of 1977
Bureau of Mines	U.S. Bureau of Mines
MINER Act	Mine Improvement and New Emergency Response Act of 2006
OSH Act	Occupational Safety and Health Act
S-MINER Act	Supplemental Mine Improvement and New Emergency Response Act of 2007
The Bureau	U.S. Bureau of Mines

One Hundred Years of Federal Mining Safety and Health Research

John A. Breslin, Ph.D.
Physical Scientist
Office of Mine Safety and Health Research, NIOSH

1. Introduction

This publication provides an historical overview of research undertaken by the U.S. federal government over the last 100 years to improve the health and safety of our nation's miners. Federal research efforts began with the establishment of the U.S. Bureau of Mines (USBM, or the Bureau) in 1910. They have continued over the past century, even after the Bureau's closure in 1996. It is hoped that this publication will give the reader an appreciation for the work of mining health and safety researchers over the past century, and of the miners served by this research.

Although not a comprehensive history, this report highlights the key organizational changes made within the Bureau and the federal government that affected mining safety and health research. Some mention is also made of Bureau research not directly related to mining health and safety. Note that the work classified as safety and health research has varied over the last century. For example, before 1970 the Bureau separated mining research and safety and health research. At that time, mining research included ground control and methane drainage because they had direct implications for mine safety. Explosives research was also reported separately from health and safety, even though permissible explosives research was obviously related to safety. Therefore, where budget figures from the Bureau are mentioned in this report, it is often not possible to separate safety and health research from other types of research.

This history is generally chronological, with a few exceptions. Because of their prominence and direct links to Bureau research, most of the Public Health Service mining studies completed before 1970, when NIOSH was established, will be discussed in a single section, even though the work spanned many decades. Likewise, for convenience, the brief histories of the Bureau's mining research centers are located in one section.

2. The Beginning: 1910-1925

2.1 Early Legislation

In 1865, Senator William Morris Stewart of Nevada introduced the first bill to establish a federal mining bureau. Congress introduced further bills during the 1800s, mostly initiated by members of Congress from western states that mined precious metals. Eastern states seemed to have less interest in a mining bureau, but some interest in establishing mining schools and experiment stations similar to those which the federal government had fostered for agriculture [USBM 1925]. At its first annual convention in 1896, the American Mining Congress proposed the establishment of a National Department of Mines in the President's cabinet [Powell 1922].

The origins of the USBM were in the U.S. Geological Survey (USGS). Congress passed a law in 1904 authorizing the USGS to analyze and test coal and lignite at the Louisiana Purchase Exposition in St. Louis, MO. This work was overseen by the expert in charge, Dr. Joseph A. Holmes, and continued at St. Louis until 1907, when the closing of the exposition transferred all work to the Jamestown Exposition at Norfolk, VA. The technological branch of the USGS was also formed in 1907, and coal testing work was placed in this branch. In 1908 the work on coal testing was transferred to the arsenal grounds of the War Department in the Lawrenceville section of Pittsburgh, PA.

On May 22, 1908, Congress passed an appropriations act authorizing investigations of the causes of mine explosions. Next, a mine accidents division of the technological branch of the USGS was formed. Mine explosions research commenced at Pittsburgh in November 1908. Initially, explosives were tested to determine those that could be safely used in coal mining. The first list of permissible explosives was published in 1909. The technological branch also investigated mine explosions directly at the sites where they had occurred. Also, the USGS established stations equipped with mine rescue apparatus at Urbana, IL, Knoxville, TN, and Seattle, WA.

The USGS technological branch had two purposes: to increase efficiency in fuel and mineral production and to increase mining safety. The western metal mining states were highly interested in a federal mining bureau, while other states held a general interest in improving mining safety. Many organizations actively promoted the creation of a federal mining agency, including the major mining industry association at that time, the American Mining Congress.

2.2 Early Mine Disasters

To understand the need for a federally funded mine bureau 100 years ago, one has to grasp the scope and context of the disasters at the time. In the early 20th century, many of the workers in American coal mines were boys as young as 10 years old. Therefore, many victims of coal mining disasters were very young boys (Figure 1). Further, inherent dangers existed in the level of available technology at the time, as represented by miners wearing a cloth cap with an open flame oil cap lamp (Figure 2).

Figure 1. Young boys working in tipple, a type of rudimentary coal screening plant.

Figure 2. Child miner wearing cloth cap with open flame oil cap lamp, circa 1900.

December 1907 saw a series of coal mine explosions that killed more than 600 miners. The first explosion, in the Naomi Mine of the United Coal Company near Belle Vernon, PA, on December 1, 1907, killed 34 miners. A coroner's jury determined that the miners died as a result of a gas and dust explosion. The gas accumulated because of insufficient ventilation and was ignited from arcing electric wires or an open light.

Five days later, on the morning of December 6, 1907, explosions occurred at the No. 6 and No. 8 mines at Monongah, WV (see Figure 3). The mines, connected to each other underground and located six miles south of the town of Fairmont, WV, were relatively modern for the time and used electrical equipment for coal cutting and haulage. Four miners emerged from the mines shortly after the explosions and an immediate rescue effort began, but the toxic gases left in the mine slowed the rescue efforts. About five hours after the explosions a surviving miner, Peter Urban, was found underground next to the body of his brother (see Figure 4). Although the rescue effort continued for several more days, no other survivors were found. Three hundred and sixty-two miners died in the mine, resulting in a makeshift outdoor morgue to accommodate the dead bodies (see Figure 5), and making the Monongah explosion the worst mine disaster in U.S. history.

Figure 3. Headline from the 1907 Monongah coal mine disaster.

Figure 4. Peter Urban, Monongah Mine disaster survivor.

Figure 5. Outdoor morgue following Monongah explosion.

A third major disaster on December 19, 1907—a gas and dust explosion—killed 239 miners at the Darr Mine near Jacobs Creek in southwestern Pennsylvania. Many of the miners killed were Hungarian immigrants. A few were former workers from the Naomi Mine that had closed because of the explosion there earlier in the month. A follow-up inquiry determined that the explosion resulted from miners carrying open lamps in an area cordoned off the previous day by the fire boss. The mine's owner, the Pittsburgh Coal Company, was not held responsible, but abandoned the use of open lamps after the disaster. The Darr Mine explosion remains the worst coal mining disaster in Pennsylvania history (see Figure 6).

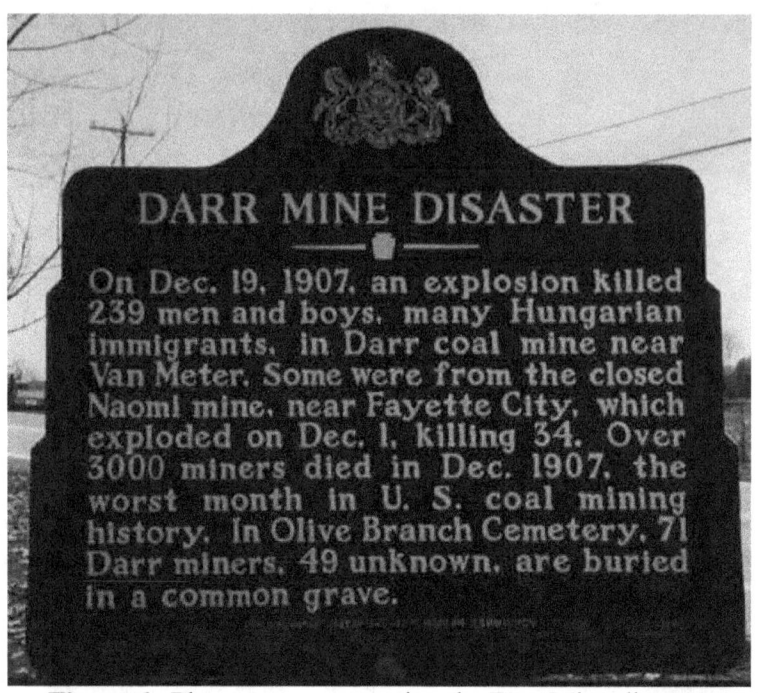

Figure 6. Plaque commemorating the Darr Mine disaster.

The second worst mining disaster in U.S. history, after the Monongah disaster, occurred on Saturday, November 13, 1909, at the Cherry Coal Mine in Cherry, IL. Nearly 500 men and boys were working in the mine that day. At about 1 p.m. a car full of hay, brought down to the second level of the mine to feed the mules, was accidentally ignited by a portable kerosene lamp, in use because of an electrical failure a few days earlier. The fire made its way up the timbers and stairs of the mine shaft, eventually burning the surface building that housed the mine fan.

As the fire continued, many miners were saved by escaping up the hoist shaft. Twelve men made six successful rescue trips up and down this shaft, but on the seventh trip, confusing signals caused the hoist cage carrying the rescuers to be lowered down the shaft instead of being raised. As a result, the rescuers and the miners with them on the cage were all killed.

Over the next few days, mine rescuers entered the mine wearing newly developed air supply apparatus (see Figure 7). On the eighth day after the fire, the rescuers met eight miners who had barricaded themselves deep in the mine and were trying to find their way out. They directed the rescuers on to the barricaded area where 13 other trapped miners were found alive. A total of 259 miners and rescuers died in the mine fire. All but three of their bodies were eventually recovered from the mine.

Figure 7. Rescuers (nicknamed "helmet men" because of the devices they wore) equipped with newly developed air supply apparatus at the Cherry Mine disaster.

2.3 The Origins of the USBM and the Organic Act

As a response to tragedies such as those at the Monongah Mine and Cherry Coal Mine, the sentiment in favor of government regulation increased further and led Congress to pass an act (36 Stat. 369) creating the USBM in the Department of the Interior. The act that established the USBM was approved by Congress on May 16, 1910, and became effective on July 1, 1910. The general aim of the Bureau under the original Organic Act of 1910 was "to increase health, safety, economy, and efficiency in the mining, quarrying, metallurgical, and miscellaneous mineral industries of the country." The words "safety" and "efficiency" were prominent on the seal of the USBM (see Figure 8).

In creating the Bureau, the Organic Act also transferred the employees and functions of the technological branch of the USGS. Originally headquartered at 8th and G Streets NW in Washington, DC, the Bureau began with 124 employees and expanded to 230 in its first year. The first mining experiment station resided at 40th and Butler Streets in Pittsburgh. The Bureau also maintained six mine safety stations and seven mine rescue cars distributed among coal fields throughout the country. Following his tenure with the USGS technological branch, Dr. Joseph A. Holmes, who had also formerly been the state geologist for North Carolina, was appointed as the first director of the Bureau on September 3, 1910 (see Figure 9).

Figure 8. Official seal of the United States Bureau of Mines, established in 1910.

Figure 9. Photo of Dr. Joseph A. Holmes.

Originally, the Bureau had two divisions—one for studying mine explosions and the other for analysis and testing of mineral fuels. The budget for the first year of operations was $502,200 (about $11 million in 2009 dollars), with $310,000 of that earmarked for mining safety studies. The annual salary of the Director of the Bureau was set by Congress at $6,000.

Beginning with 1911, the Bureau Director produced an annual report for each fiscal year. These annual reports continued after the Bureau was transferred to the Department of Commerce in 1925, then transferred back to the Department of the Interior in 1934. After that year, the annual reports for the Bureau were absorbed into section of longer reports of the Secretary of the Interior.

In 1912, both the Bureau Director, Dr. Joseph A. Holmes, and the assistant to the director, Van H. Manning, were based at the Bureau headquarters in Washington, DC. The mining experiment station at Pittsburgh was headed by the engineer-in-charge, H. M. Wilson, assisted by the chief mining engineer, G. S. Rice, the head of the mining division at Pittsburgh. This division was responsible for health and safety investigations, miner training, and inspection of mines in the territory of Alaska.

2.3.1 Initial Safety Research

In 1910, the United States had more than 700,000 coal miners working in 15,000 coal mines, producing about 500 million tons of coal annually. Much of the initial work by the Bureau was educational. During its first year the Bureau trained over 50,000 miners in first aid, mine rescue, and fire-fighting. The Bureau's mine rescue cars were used to train miners and mine rescue teams throughout the country (see Figure 10).

Figure 10. Dr. Joseph A. Holmes and miners being trained in front of mine rescue car, 1913.

When the Bureau was established in 1910 the fatality rate in US coal mines was 3.91 per 1,000 miners employed. By comparison, fatality rates were 1.70 per 1000 in England and 1.98 in Prussia. In U.S. metal mines, the fatality rate was also above 3 per 1,000 employees.

In response to these troubling fatality rates, the Bureau's initial mine safety research focused in these areas:

- Explosibility of mine gas and coal dust and prevention of explosions and fires;
- Safety of explosives used in mines;
- Electrical safety in mines;
- Safety of mine lights and their use as gas detectors;
- Emergency breathing apparatus.

Most of this initial research was specific to coal mines, but the Bureau also actively sought funds to study safety and health problems in metal mines [USBM 1911].

Safety explosives research begun in 1908 under the USGS continued under the Bureau. The Bureau's second annual report [USBM 1912] noted the increasing use of new types of explosives beginning to replace black powder in coal mines (see Figure 11). These newer explosives were safer because they generated shorter and cooler flames, and thus were less prone to ignite any nearby methane gas. Those meeting Bureau safety criteria were designated as permissible explosives. In 1912, the Bureau conducted over 10,000 explosives tests and added 32 permissible explosives for use in gassy or dusty coal mines [USBM 1912]. The use of permissible explosives reduced both the occurrence of mine disasters and the number of individual accidents causing injury to miners.

Figure 11. Miner preparing explosive shot in coal mine.

Under the USGS, the Bureau also began to compile information on state laws and regulations pertaining to mine safety. This information was passed on to the states to improve mine safety legislation and directly to mine operators and miners to increase safety in coal mines. The Bureau sent engineers to visit every mine that had suffered a significant explosion or fire to investigate the cause and offer assistance where needed. These investigations were completed "in cooperation with or the approval of the state or mine officials" [USBM 1912]. At that time, the Bureau had no statutory authority to enter and investigate mines without permission.

2.4 The USBM and Miner Health

In response to diseases that emerged among the mining population, the USBM began to investigate miner diseases concentrating on the lung illness then called "miner's consumption" or "miner's phthisis." These diseases, which we now call pneumoconiosis or silicosis, often occurred together with tuberculosis. The Bureau advocated methods to reduce or completely eliminate the disease from mining camps. The Bureau's investigation of lung disease in miners was done in cooperation with the Public Health Service, beginning a history of partnership between the two agencies.

The Bureau's first major study of miners' health was on the prevalence of silicosis among metal miners in the lead-zinc mines near Joplin, Missouri. Edwin Higgins, a Bureau mining engineer, and by Dr. A. J. Lanza, an assistant surgeon on detail to the Bureau from the Public Health Service, led the study. This study produced findings on the causes and methods of abating dust in mine air, the chemical and physical characteristics of dust, and the concentrations of dust in mine air. Notably, the investigators also developed an airborne dust sampler for this study (see Figure 12). Powered by the breathing of the person taking the sample, the airborne dust sampler used granulated sugar as the filtering medium to collect the dust. From the resulting measurements, the investigators concluded that by using available dust controls, it was possible to control airborne dust in mines below 1 milligram per 100 liters of air.

Figure 12. Early airborne dust sampler developed by the USBM, circa 1915.

Seven hundred twenty miners were given medical exams as part of this first Bureau study. Of these, 330 or 45.7 percent were found to have miners' consumption alone, and an additional 14.7 percent had miners' consumption along with tuberculosis infection [Higgins et al. 1917].

2.5 The USBM Opens an Experimental Mine near Pittsburgh

In its first year, the USBM planned an experimental mine for testing of coal dust explosions. A suitable location for this mine was found on the property of the Pittsburgh Coal Company, near Bruceton, PA, about 13 miles south of Pittsburgh (see Figure 13). The Bureau began to lease the mine property annually from its owner. In December 1910, drift openings were begun into the outcrop of the Pittsburgh coal seam. By October 1911, two parallel entries were driven more than 700 feet into the seam. The entries were 9 feet wide with a 41-foot pillar between them, and crosscuts were made at 200-foot intervals.

Figure 13. Early view of the USBM's Bruceton, PA, facility.

The first public test of a coal dust explosion in air containing no inflammable gas was made in the Bruceton experimental mine on October 30, 1911. This demonstration in a full-scale underground mine helped prove that coal dust could explode in mines without the presence of methane gas [Rice et al. 1913]. The Bruceton mine was expanded in 1912 with two main entries 980 feet into the outcrop in the Pittsburgh coal seam. The primary use for the mine was to study full coal dust explosions, which was not possible under smaller laboratory conditions or in a scale model of a mine. Early use of the mine also included testing of gasoline locomotives, mining machinery, explosives, electrical equipment, and ventilation methods.

The Bureau organized a national mine safety demonstration at Forbes Field in Pittsburgh on October 30, 1911. The demonstration included exhibits of the Bureau's research on mine safety, mine rescue, and first aid, with coal dust explosions at the experimental mine at Bruceton. About 15,000 people attended, including the President of the United States, William H. Taft (see Figure 14). More than 1,200 people toured the Bruceton mine and witnessed the live coal dust explosion demonstration (see Figure 15).

Figure 14. President William H. Taft and Dr. Joseph A. Holmes at national mine safety demonstration, Pittsburgh, 1911.

Figure 15. Public demonstration of coal dust explosion at Bruceton experimental mine.

From its early years, the Bureau placed great emphasis on publicizing its research results, public exhibitions on mine safety and technology, and educating miners and mine managers (see Figures 16-18).

Figure 16. U.S. Bureau of Mines exhibit at San Francisco Exposition, CA, 1915.

Figure 17. Mine rescue exhibition at Forbes Field, Pittsburgh, PA, circa 1918.

Figure1 18. USBM exhibit at Century of Progress exhibition in Chicago, IL, 1933.

2.6 Expansion of the USBM Mission

By 1912, the USBM recognized that most fatalities in coal mine explosions were caused by the release of toxic mine gases rather than the explosions themselves. In emergencies, some rescued miners had successfully barricaded themselves in isolated areas of the mine where the air was breathable, and thus were not affected by the gases. In response, the Bureau's second annual report [USBM 1912] described refuge chambers that might be built in mines for use by miners after an explosion. The chambers would be simple in construction and contain food, water, and first aid materials. Several Bureau engineers advocated that they be made available in all coal mines. The report also introduced the possibility of providing a protected telephone for communication and boreholes to the chambers in mines closest to the surface.

The largest cause of fatalities in coal mines in 1912 was falls of roof and coal, with roof falls also identified as the greatest safety hazard in metal mines. The Bureau noted that the mining systems in use may have been key to the problem, as mines in Europe appeared to have lower fatality rates. At the time, U.S. mines used less timber for roof support and suffered from a less safe use of explosives [USBM 1912].

A new Organic Act for the Bureau was passed by Congress and signed by President Taft on February 25, 1913. This new act broadened the scope and more clearly defined the mission of the Bureau, charging it as follows:

> ... to conduct inquiries and scientific and technologic investigations concerning mining, and the preparation, treatment, and utilization of mineral substances with a view to improving health conditions, and increasing safety, efficiency, economic development, and conserving resources through prevention of waste in the mining, quarrying, metallurgical, and other mineral industries [USBM 1913].

The Bureau's purpose was described concisely at the beginning of the fourth annual report: to conduct research to "lead to increased safety, efficiency, and economy in the mining industry of the United States" [USBM 1914]. Congress required that the Bureau should not conduct investigations on behalf of private parties nor promote individual business enterprises. The Bureau policy was to be nonpartisan, favoring neither the miner nor the mine operator. Its function was to stand for the welfare of all workers in the mining industry, the industry itself, and the public.

The Bureau's fourth annual report [USBM 1914] also stated the general policy under which the Bureau operated relative to other mining stakeholders:

1. that the National Government conduct the necessary general inquiries and investigations in relation to mining industries and disseminate in such manner as may prove most effective the information obtained and the conclusions reached;
2. that each State enact needed legislation and make ample provision for the proper inspection of mining operations within its borders;

3. that the mine owners introduce improvements with a view to increasing safety and reducing waste of resources as rapidly as the practicability of such improvements is demonstrated; and
4. that the miners and mine managers cooperate both in making and in enforcing safety rules and regulations as rapidly as these are shown to be practicable.

These policies continued through the Bureau's history except for the gradual transfer of responsibility for mining regulation and enforcement from the states to the federal government as a result of subsequent legislation.

The Bureau's fifth annual report [USBM, 1915] began with a note announcing the death of Dr. Joseph A. Holmes, the first Bureau Director, in Denver on July 13, 1915, in his 56th year. Poor health had forced him to leave Washington in the summer of 1914, but he continued to direct Bureau policy until shortly before his death. Appendix A contains a short biography of Dr. Holmes.

The Bureau had accomplished much in its first five years, summarized by a brief pre-Bureau history and a listing of its accomplishments in the opening pages of the Bureau's fifth annual report [USBM, 1915]. In this report, the following categories were used to summarize early Bureau research:

- Study of coal mine explosions;
- Investigations of explosives used in coal mining;
- Investigation of use of electricity in mining;
- Investigation of mine lighting;
- Use of rescue and first aid apparatus;
- Study of safety and health problems in metal mines;
- Inquiries as to waste of metals, ores, and minerals in treatment and use;
- Petroleum investigations;
- Compilation of mining laws and regulations.

Note that seven of these nine major headings relate to safety and health of miners, particularly in coal mines. Further, it is noteworthy that as of mid-1914 nearly 100 men had been rescued from mines by Bureau employees, with many more saved by Bureau-trained rescuers [USBM 1914]. One method employed by rescuers was the use of a collapsible cage to carry miners to the surface (see Figure 19).

Additional highlights of the Bureau's work in its first five years included:

- Coal dust was proved to greatly increase the magnitude of mine explosions over those involving methane alone.
- Use of permissible explosives tested by the Bureau had reduced the danger of coal mine explosions caused by the use of explosives.
- Bureau work had led manufacturers to produce safer electrical apparatus and cap lamps for miner use in gassy coal mines. Recommendations of the Bureau led several states to enact stricter laws regarding electrical equipment.

- Bureau demonstration of rescue apparatus and training of miners in first aid had led to the rescue of 200 men after mine disasters and strengthened the movement for increasing safety in all industries.
- A Bureau investigation in cooperation with the Public Health Service in Missouri had revealed an excessively high mortality from tuberculosis and the importance of siliceous dust in mine air as a causative agent.
- The Bureau had begun compiling statistics on mine accidents and fatalities based on the data provided by the states. Statistics were also collected on safety in metallurgical plants.

Figure 19. Collapsible cage used for mine rescue.

2.7 Expansion of the USBM in Pittsburgh

In 1915, the Bureau began construction of new buildings for its experiment station in Pittsburgh (see Figure 20). Adjacent to the then campus of the Carnegie Institute of Technology (now part of Carnegie-Mellon University), the new buildings were completed in 1917 and located on Forbes Avenue in Pittsburgh (see Figure 21). Also in 1915, the Bureau began a test in a commercial mine of the Pittsburgh Coal Company to determine the efficiency of rock dust to render coal dust inert and prevent coal dust explosions. It was hoped that this method might be more efficient than the use of water for the same purpose on a large scale [USBM 1915].

Figure 20. The USBM's Forbes Avenue facility under construction, Pittsburgh, PA, circa 1915.

Figure 21. The USBM's completed Forbes Avenue building, Pittsburgh, PA, 1917.

2.8 Early Statistics on Mining Fatalities and Production

In 1916, the Bureau published a report documenting mine safety trends from the earliest published reports to that time [Fay 1916]. In particular, this report compiled accident statistics from the states that had inspected mines from 1870 to 1914. Pennsylvania had fatality records reaching back to 1870 for anthracite mines and to 1877 for bituminous mines. By the time the Bureau was established in 1910, almost 90 percent of coal production came from states with mine inspectors who reported coal mine fatalities from their states.

The Bureau's report calculated that from 1839 to 1914, the total number of coal mining fatalities reported was 53,078 [Fay 1916]. The report also included USGS data on coal mine production and employment. Based on these data, the Bureau estimated that the number of fatalities in all U.S. coal mines up to that time totaled 61,187.

Complete records for coal mine employment first became available in 1889, when the number of men employed in U.S. coal mines was 311,717. In that year, 90.8 percent of miners and 90.5 percent of coal production came from states with mine inspectors who reported mining fatalities.

The fatality rate per million tons mined was 13.47 in 1870 and gradually reduced to 4.78 by 1914. The average amount of coal mined per man per year increased from 440 tons in 1870 to 673 tons in 1914. In 1914, the productivity rate was 3.20 tons mined per man per day [Fay 1916].

Falls of roof and pillar coal were the largest cause of coal mining fatalities, taking the lives of 23,260 miners from 1870 to 1913. The fatality rate from this cause increased from 1.60 per 1000 men employed in 1870 to 1.69 in 1913, with little variation in the rate during those years [Fay 1916].

Gas and dust explosions were another major cause of fatalities. The lowest rate was in 1887 (about 4 percent of fatalities), and the worst year was 1907 (almost 30 percent). The fatality rate in 1907 from gas and dust explosions was 1.42 fatalities per 1,000 men employed [Fay 1916]. Thus, even in 1907, the worst year for coal mine explosions, the fatality rate from falls of roof and pillar coal was still higher by comparison.

In the 1870s, fatality rates declined, then leveled off between 1880 and 1897, before increasing from 1897 to 1907 [Fay 1916]. This increase can be attributed to the mines becoming deeper and to the higher average number working in each mine. Also, the influx of foreign laborers, many of whom had no mining experience and could not communicate well in English, was a potential factor. A gradual decrease in the fatality rate between 1907 and 1914 was attributed to more stringent state mining laws, Bureau educational campaigns on mine accident prevention, and technologies such as permissible explosives, improved safety lamps, and precautions taken to render coal dust inert [Fay 1916].

2.9 The USBM and WW I

In its early years, the USBM studied poisonous and explosive gases. Therefore, in February 1917, as the U.S. was about to enter World War I, the Bureau assisted the War Department in studying poison gases and gas masks, until the effort was transferred to the Army in June 1918.

Other Bureau contributions to the war effort included development of aviation fuels, recovery of helium for airships, studies on explosives production, and investigations on the availability of scarce minerals for war use. The Bureau was also given the authority during World War I to regulate the production and use of explosives. Following the war, the Bureau continued to develop protective equipment against mine and industrial gases and dusts, including gas masks effective against carbon monoxide and respirators for use in oxygen-deficient atmospheres (see Figure 22).

In 1919, the Bureau was asked to study the effects of motor vehicle exhaust gases in the confined space of tunnels, in connection with the planning of the first long vehicular tunnel under the Hudson River between New Jersey and New York City. Research at the Bruceton Experimental Mine established ventilation requirements for maintaining carbon monoxide from automobile exhaust at maximum allowable levels. This was accomplished within the Experimental Mine through construction of an oval track, whereby test cars were run around the track while observing the effects of temperature, humidity, airflow rate, smoke, and exhaust gases on the drivers (see Figure 23).

A carbon monoxide detector installed in the Liberty Tunnels in Pittsburgh was one of the first of a long series of atmospheric monitors and detectors developed by the Bureau at Pittsburgh. The Bureau also undertook studies on first aid treatment for carbon monoxide poisoning victims. This research resulted in the ventilation system used in the Holland Tunnel in the early 1920s and in other major tunnels built later [CRS 1976]. The Bureau's interest in this area continued, and in 1932 it participated in a health study with the Port of New York Authority involving 175 policemen on duty at the Holland Tunnel.

Figure 22. Miners wearing the Gibbs mine rescue breathing apparatus, 1918.

Figure 23. Ford Model T in the Bruceton Experimental Mine, 1921.

3. Hard Times and Maturation: 1925-1969

3.1 USBM Transferred to the Department of Commerce

An executive order by President Coolidge transferred the USBM from the Department of the Interior to the Department of Commerce effective July 1, 1925. By that time, the Bureau had 830 employees, of which 789 held Secretary's appointments. Of these 789 employees, 211 were located in Washington, DC, 265 were based in Pittsburgh, PA, and the other 313 were at other field locations. Three hundred seven were classified as engineers, chemists, or similar technical occupations. By 1925 the Bureau had 12 experiment stations [USBM 1925].

3.1.1 Explosives Research

Throughout the 1920s and 1930s, the Bureau continued to study explosives at the Bruceton Experimental Mine (see Figure 24), including tests on the amounts of explosives and stemming needed to protect against ignitions of mine gas or coal dust and the effects of blasting on mine roof.

One of the Bureau's technologic branches was the Explosives Division. Under the USGS, the government had begun explosives testing in 1908. In 1907, the fatality rate in coal mining was 1.687 per thousand miners employed; by 1929, the fatality rate had been reduced to 0.432 per thousand miners. A large factor in this improvement was the transition to using permissible explosives. In 1907, only about 1 percent of explosives used were of the short flame (permissible) type, whereas by 1929 about 50 percent of explosives used were permissible [USBM 1932].

Figure 24. Explosion test at the Bruceton Experimental Mine, 1927.

Beginning in 1922, the Bureau studied the breaking of ore and coal by explosives to determine optimum blasting patterns and timing for efficient and safe blasting in underground mines. Much of this work was completed at the Bruceton Experimental Mine as well as at two new experimental mines established by the Bureau at Mount Weather, VA, in 1936 and at Rifle, CO, in 1945. Research at Mount Weather included studies on drilling and blasting and on methods for setting industrial diamonds in drills to increase drilling speed and efficiency. The experimental mine at Rifle studied efficient methods for mining of very thick deposits of oil shale in western Colorado.

3.1.2 Rock Dusting Research

The possibility of using rock dust to prevent coal dust explosions was recognized as early as 1911 [Rice 1911]. Building on this work, the Bureau began an extensive series of tests, and by 1922 these tests in the Bruceton Experimental Mine verified that rock dust could help prevent dust explosions [Rice et al. 1922]. Some commercial coal mines were already using rock dust as a preventive measure as early as 1912 and the practice spread, with manufacturers producing commercial rock dusting machines by 1930.

The practice of rock dusting spread slowly. By 1930, 8.4 percent of underground bituminous coal mines were using rock dust, and these mines accounted for 33.2 percent of underground coal production. By 1940, the underground coal mines being rock dusted increased to only 9.1 percent; however, these mines were responsible for 55.7 percent of underground production. Thus, the large high production mines were much more likely to be using rock dust.

In the 1940s, some states required rock dust use in underground coal mines, but many mines that used rock dusting were believed to be doing so ineffectively. By 1950, the figures had increased to 29.7 percent of mines producing 84.5 percent of the underground coal. An explosion at the Orient No. 2 mine in 1951 was believed to have been caused in part by ineffective rock dusting, and the Federal Coal Mine Safety Act of 1952 introduced federal regulations requiring and giving specifications for the application of rock dust to coal mine surfaces. This law required that enough rock dust be added to coal mine surfaces to maintain a minimum of 65 percent incombustible material [Humphrey 1960].

3.1.3 Safety Inspection

The Bureau performed limited safety inspections during the 1920s and 30s. For example, in 1932 the Safety Division of the Bureau prepared 204 confidential reports on safety inspections of mines and other work sites made by its field staff. The inspections included coal, metal and nonmetal mines, quarries, and petroleum plants. Of these reports, only 84 were given to operators because of their confidential information, and none were published [USBM 1932]. At this time, the Bureau had no authority to inspect mines without the mine operator's permission.

The Bureau's 1933 annual report [USBM, 1933] noted that 1,463 persons were killed in coal mines in 1931, and another 1,166 in 1932. This compared to an average number of coal mining fatalities of 2,409 annually over the previous 25 years. In 1931, only 3.31 miners were killed per million tons of coal produced in the U.S.—the lowest rate up to that point in the 20th century.

3.2 USBM Transferred Back to the Department of the Interior

In 1934, the USBM was transferred back to the Department of the Interior from the Department of Commerce. At that time, the Bureau was organized in four branches: Technologic, Health and Safety, Economics, and Administrative. The Health and Safety Branch included three divisions: Health, Safety, and Demographic. The administrative headquarters was in Washington, DC, but most staff worked at the experiment stations and field offices scattered throughout the country's mining and petroleum districts.

Most Bureau research other than health and safety was in the Technologic Branch, but this branch's Explosives Division and the Mining Division also did safety research. The Mining Division studied mine ventilation, rock mechanics, and ground control practices to reduce fatalities from roof falls. Unfortunately, roof fall research was suspended in July 1933 because of a shortage of funding. In 1934, most of the research in the Technologic Branch resided within the 11 experiment stations at Pittsburgh, PA, New Brunswick, NJ, Minneapolis, MN, Tuscaloosa, AL, Rolla, MO, Bartlesville, OK, Reno, NV, Salt Lake City, UT, Tucson, AZ, Berkeley, CA, and Seattle, WA.

Within the Health and Safety Branch, Health Division research was suspended in July 1933 because of the shortage of funds. The Safety Division continued to supply safety training and mine rescue and recovery, and to perform research on mine accident prevention. In 1934, Safety Division field staff consisted of 22 engineers and 23 safety instructors located at 10 safety stations in mining areas. Over the previous seven years, they had trained about 520,000 persons in first aid or mine rescue. The Demographic Division collected statistics on mine safety and production of explosives; i.e. it conducted federal mine safety surveillance [DOI 1934, DOI 1935].

3.3 Impact of the Great Depression on the USBM

Along with the rest of the country, the USBM fell upon very hard times in the early years of the Great Depression. Scott Turner, the then Bureau Director, gave a paper to the American Mining Congress in late 1933 describing the effects of budget reductions on the Bureau [Turner, 1934]. The budget of the Bureau had been reduced from a peak of $3,444,595 in 1929 to $1,860,325 ($31 million in 2009 dollars) for 1933. For 1934, the budget appropriated by Congress for the Bureau was reduced to $1,514,300, and the Department of Commerce reduced the amount available to the Bureau further to only $1,100,000 (see Table 1).

As a result of these budget reductions, the Bureau dismissed almost 200 employees at one time in July 1933. The total staff reduction was from 755 employees in March 1933 to 501 employees in December 1933, or about one-third of the staff in less than one year. The remaining employees suffered a 15 percent salary reduction and an enforced leave without pay for 15 days due to the impounding of $413,300 from the Bureau's budget by the Department of Commerce [DOI 1935]. Other budget restrictions included the decommissioning of 9 of the 11 mine rescue railroad cars, which had to be stored on sidings, and closure of field offices in Laramie, WY, Shreveport, LA, Phoenix, AZ, Evansville, IN, and Kansas City, MO.

With the cuts forced by the Great Depression, all research on falls of mine roof, explosives, and mechanical equipment in underground mines was forced to be terminated. The work of the Bureau's Health Division was also terminated. The Bureau had to ask the Public Health Service to take back to its own payroll the officers who had been assigned to the Bureau. This ended for a time the close cooperation between the Bureau and the Public Health Service on improving the health of miners [DOI 1935].

3.4 The USBM and WW II

The Federal Coal Mine Inspection and Investigation Act of 1941 authorized the USBM to enter and inspect coal mines to observe hazards and recommend safety improvements. Previously, the permission of the mine operator had been necessary to enter a coal mine. In the first 4½ years under the Act, the mining industry complied voluntarily with approximately 200,000 or 35 percent of the recommendations made by Bureau coal mine inspectors. By 1946, the Bureau enjoyed a field force of 157 coal mine inspectors and 10 explosives or electrical engineers. The inspection force grew to 200 by 1948. Meanwhile, some of the states had modified their mining laws to partially parallel the Bureau's inspection standards and procedures.

During the war years of 1942-45, 5,314 lives were lost in coal mines while 2,608,000,000 tons of coal was produced—a fatality rate of 2.04 deaths per million tons of coal. This compared with a total of 7,502 deaths in 1916-18 at the time of World War I, during which 1,919,000,000 tons were produced—a fatality rate of 3.91 deaths per million tons of coal. Thus the fatality rate had been reduced by almost 50 percent between the two world wars.

After World War II, the Bureau performed some preliminary testing of air quality in mines using diesel mine locomotives. It was concluded that "with reasonably good ventilation and approved operating procedures and maintenance, diesel engines of suitable design may be used in mining operations without creating hazardous or unduly objectionable conditions in the air of working places" [DOI 1946].

In 1948, more than 30,848 mining industry employees were instructed in first aid and mine rescue work, bringing the total receiving training in these areas to 1,705,489 since the Bureau was established in 1910. In that year, the Bureau initiated a course for mine safety committeemen in cooperation with the United Mine Workers Union. The course was given to 2,000 out of the 12,000 mine safety committeemen, as well as to 1,500 other miners [DOI 1948].

3.5 Significant USBM Contributions Outside of Mining

3.5.1 Explosives and Flammability Research

USBM research on explosives expanded during World War II into an extensive fundamental research program on explosives and propellants. This work was done in cooperation with the National Defense Research Committee. Coal dust explosions research expanded after the war into studies of explosions of other industrial and agricultural dusts such as aluminum powders and grain dusts.

As a result of this effort, Bureau explosives researchers discovered new methods for measuring the explosives detonation rates and the temperature in detonation waves. Researchers also investigated the explosibility of ammonium nitrate, which had caused the Texas City, TX, disaster in 1947, claiming 581 lives [CRS 1976]. They discovered that this material could be detonated by heat alone under certain conditions of confinement.

Building on its expertise in explosives research, the Bureau also studied the ignition hazards of static electricity in hospital operating rooms where anesthetic gases were used. Researchers identified possible hazards and recommended practices to avoid explosion problems. This research was useful not only for understanding and preventing coal mine explosions, but was applicable to the prevention of explosions in airplanes, submarines, and industrial plants [CRS 1976].

Over the course of several decades, significant fundamental research was undertaken by the Bureau on the nature and characteristics of flame. Under a cooperative agreement between the Bureau and the Safety in Mines Research Board of Great Britain, Dr. H.F. Coward of Sheffield, England, was detailed to the Bureau's Mines Experiment Station at Pittsburgh in April 1925. There he worked with G.W. Jones and other Bureau researchers on conditions for the initiation and propagation of flame in different mixtures of gases under various conditions. Their results were compiled in several Bureau Bulletins, the last of which was published in 1952.

Dr. Coward eventually became the Director of Safety in Mines Research, Ministry of Fuel and Power, Great Britain. Certainly the work of Dr. Coward and his collaborators was important in preventing gas explosions in coal mines, but the results were also widely applicable in metallurgical, petroleum, gas production, and other industries [Coward and Jones 1952]. The collaboration of the Bureau with the Safety in Mines Research Board continued for many years and also included advancements in mine rescue apparatus, fundamental studies of flammability of coal dust, and rock dusting to prevent coal mine explosions.

Bureau technology for the coal mining industry was also applied to a wide range of potentially explosive and flammable materials of interest to other industries and government agencies, including the Department of Defense, the Air Force, the Department of Transportation, the Coast Guard, and the National Aeronautics and Space Administration (NASA). The Bureau's Bruceton Experimental Mine did preliminary work on the explosive trigger for the atomic bomb—specifically, the first experimental work on cylindrical implosions.

Bureau scientists also developed methods for determining flammability hazards of combustible materials. Standard tests to determine auto-ignition temperatures, flammability limits, and minimum ignition energies, developed by Bureau personnel, were adopted by the American Society for Testing and Materials. Fundamental research investigated the manner in which flames and explosions originate, propagate, and are quenched.

In the second half of the 20th century, the Bureau's Bruceton Experimental Mine made major contributions to the formulation of new military explosives and to the development and safe use of fuels and fluids for conventional aircraft, jet engines, and space vehicles. Projects for the military and NASA included research for the destruction of earthen tunnels in Vietnam; the effect of the extraterrestrial environment on Apollo attitude control engines; investigations on the safe handling of hazardous materials on oil tankers; the storage and transport of liquid natural gas, liquid hydrogen, and liquid chlorine; the safest method for ejecting burning fuel from a space capsule at an altitude of 100,000 feet; and the effect of extraterrestrial atmospheres on the performance of explosives [Tuchman and Brinkley 1990]. Dr. Robert W. Van Dolah of the Bureau was one of an eight-member board that investigated the disastrous fire aboard an Apollo space capsule in 1967 in which three astronauts were killed.

3.5.2 Metals Research

Titanium metal, which is high strength and light weight, could not be produced in practical quantities before World War II. Therefore, in 1943, the Bureau began research efforts to produce the metal in large quantity. Researchers developed an improved process based on one originally created by Dr. Wilhelm Kroll, who had fled the Nazis to America. The Kroll process was enhanced and in 1944 the Bureau began to produce titanium metal at a pilot plant in Boulder City, NV. Further research by the Bureau perfected the process so that production of titanium metal by the private sector was begun in 1948.

Bureau research on zirconium, with desirable properties similar to titanium, took place at its laboratory at Albany, OR, with the goal of producing practical quantities of zirconium by way of a process adapted from that used for titanium. As a result, beginning in 1945, the Bureau produced most of the national supply of metallic zirconium. This metal was used by the U.S. Atomic Energy Commission in atomic power applications. Eventually this process was taken over by private companies to produce zirconium metal in quantity [CRS 1976].

3.6 Mechanization of Coal Mining over the Course of the Century

Underground coal mining production methods evolved several times throughout the 20th century. At the beginning of the century, traditional pick mining and blasting, with hand loading of the broken coal, was the norm (see Figure 25). By 1930, undercutting machines had largely replaced hand picks for undercutting and mechanical loading had greatly increased (see Figure 26). Conventional mining by drilling and blasting was gradually replaced by continuous miners during the 50s and 60s, while longwall mining became a dominant coal mining production method during the 70s and 80s (see Figure 27).

Figure 25. Undercutting coal with a hand pick in the early 20th century.

Figure 26. Joy loading machine in 1925.

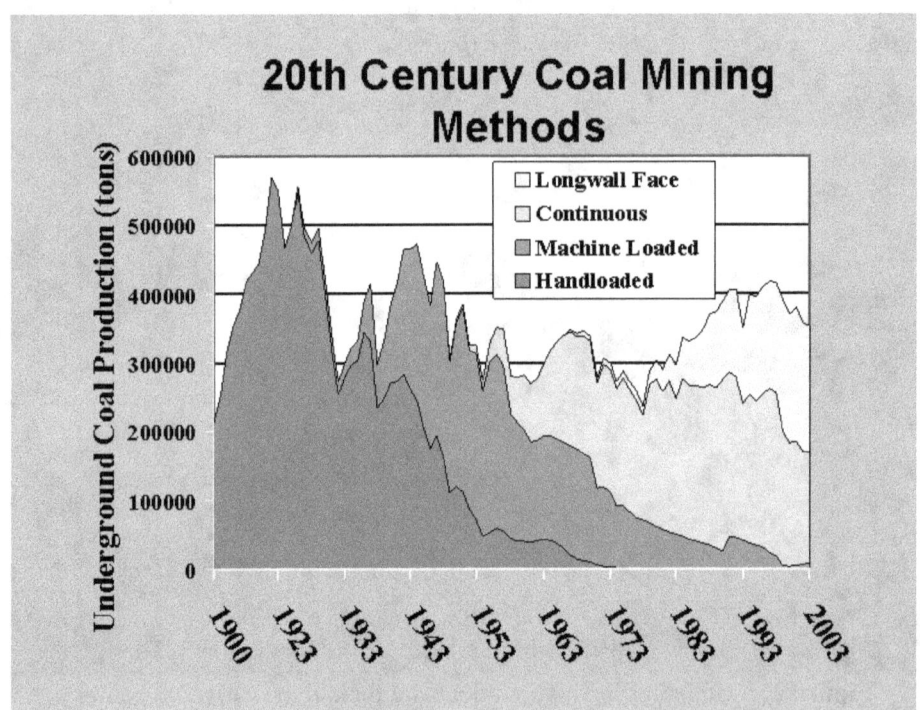

Figure 27. Coal mining production by mining method in the 20th century.

3.6.1 Roof Control

The Bureau began to devote more attention to the problems of roof control in the 1940s. Until the 1950s, wood timbers and posts were the usual method of roof support in coal mines (see Figure 28), but the Bureau saw potential in roof bolting. Roof bolts, long steel rods grouted or anchored into a drilled hole by a wedge, are 3 to 8 feet long and about 1 inch in diameter. They clamp together the roof strata to form a composite beam much stronger than the original roof strata.

Figure 28. Setting timber posts in a coal mine for roof control.

Roof bolting was first introduced in 1905 and grew popular in U.S. metal mines by the 1940s. In 1947, with the Bureau's help, Consolidation Coal Company began roof bolting trials at Consol's Mine No. 7 near Staunton, IL. As it gained confidence in the technique, the Bureau began to advocate roof bolting enthusiastically as an accident prevention measure and became involved with roof bolt trials in mines across the country. Roof bolting helped to reduce injuries from roof falls and also permitted improved mobility for mining equipment (see Figures 29-30). By 1957, the Bureau estimated that more than half of all underground bituminous coal was produced in mines that used roof bolts. Today, roof bolting is the universal primary roof support in American coal mines [CRS 1976].

Figure 29. The impact of roof bolts in the Isabella Mine, Isabella, PA, in 1949.

Figure 30. The results of conventional timbering at the Isabella Mine, Isabella, PA, in 1949.

3.7 The Federal Coal Mine Safety Act of 1952

On the evening of December 21, 1951, just before the mine was to close for the Christmas holiday, an underground explosion killed 119 miners during the second shift at the Orient No. 2 coal mine in West Frankfort, IL. The West Frankfort mine disaster was the worst of its kind since 1940 (see Figure 31).

Figure 31. John L. Lewis, President of the United Mine Workers of America, at West Frankfort disaster site, 1951.

The 119 miners killed in the West Frankfort explosion left 111 widows and 176 children. The disaster drew national attention and resulted in the passage of the Federal Coal Mine Safety Act of 1952.

The 1952 Act authorized the Bureau to enforce specific requirements aimed at the prevention of mine disasters and provided for annual inspections in certain underground coal mines. It also gave the Bureau limited enforcement authority, including power to issue violation notices and imminent danger withdrawal orders. Specifically, it authorized the assessment of civil penalties against mine operators for noncompliance with withdrawal orders or for refusing to give inspectors access to mine property. However, no provision was made for monetary penalties for noncompliance with the safety provisions. Finally, the Act required the Bureau to cooperate with mine inspection agencies in coal mining states for joint enforcement. In the first year under the Act, the following improvements were achieved [CRS 1976]:

- Systematic roof control plans were adopted in more than 2,000 coal mines;
- Use of dangerous black blasting powder was discontinued in more than 1,200 mines;
- Main ventilating fans were installed in more than 2,300 mines;
- Nearly 1,000 mines began using water to suppress dust;
- Pre-shift examinations for gas were made in more than 2,300 mines;
- More than 1,500 mines were rock dusted for the first time;
- No smoking rules were adopted in more than 1,300 mines;
- Use of open flame lights was stopped in 1,200 mines.

Relevant to the enforcement provisions of the 1952 Act, the Bureau underwent substantial reorganization in 1954-55, separating the administration of health, safety, and coal mine inspection from the scientific and technical research [CRS 1976].

3.8 Permissible Explosives Research

Historically, the adoption of permissible explosives by the coal mining industry was critical to reducing explosions in underground coal mines. However, the transition to permissible explosives required decades. In 1910, about 10 percent of explosives used in coal mines were permissible, while about 80 percent were black powder and the rest dynamite. It took until 1940 before more than 50 percent of the explosives used were permissible, although some non-permissible explosives were still frequently being used in surface coal mining. The use of black powder in underground coal mines was finally prohibited in 1966.

Bureau research also led to an increase in the authorized permissible charge limit from 1½ to 3 pounds, which helped improve mining efficiency. Other improvements in blasting were the use of water-filled stemming bags, single- and multiple-shot blasting machines, and short-delay detonators.

Bureau explosives research at Pittsburgh in the 1950s discovered that adding sodium chloride to an explosive formulation significantly reduced the probability of a methane explosion. As a result, by 1962 all active permissible explosives included approximately 10 percent sodium chloride.

Bureau research also aided in the development of slurry and emulsion explosives, which are safer in relation handling, transport, and storage. Dr. Robert W. Van Dolah, of the Bureau's Pittsburgh Laboratory, was awarded the Nitro-Nobel Gold Medal in 1967 for his theory to explain the accidental initiation of liquid explosives [Tuchman and Brinkley 1990].

4. Public Health Service Mining Studies

4.1 Joint Public Health Service and USBM Studies

In 1914, the Public Health Service established the Office of Industrial Hygiene and Sanitation to study the health problems of workers. This office eventually became the Division of Occupational Health, the Bureau of Occupational Safety and Health (BOSH), and finally the National Institute for Occupational Safety and Health (NIOSH). Its establishment in 1914 marked the beginning of the general occupational health program of the federal government. Dr. Schereschewsky, the first Chief of the Office of Industrial Hygiene and Sanitation, was based in Pittsburgh, probably because of the close relationship with the Bureau's experiment station in that city.

Joint studies of mining problems by the Public Health Service and the Bureau began in 1913 with a comprehensive study of lead and zinc mines in the Tri-State area (Kansas, Oklahoma, and Missouri). The Public Health Service and the Bureau also cooperated on later miner health studies, including uranium mining in 1950, metal mining in 1939 and 1958, diatomaceous earth mining and processing in 1953, and several more limited studies in such mining operations as mercury, copper, bentonite, phosphate, and asbestos.

4.1.1 Silicosis

Joint Public Health Service and Bureau silicosis studies began in the Tri-State Mining District in 1913. About 60 percent of the 720 miners examined were diagnosed as having "miners' consumption." Dr. Royd Sayers, one of the silicosis study investigators, remained in the Public Health Service Office of Industrial Hygiene and Sanitation until 1940, when he became Director of the Bureau. He also later served as Chief Medical Officer of the United Mine Workers Welfare and Retirement Fund.

The Vermont granite industry silicosis study was done in 1924 to 1926. In this study, dust measurements were taken with the Greenburg-Smith Impinger developed by Leonard M. Greenburg of Yale University and George W. Smith of the Bureau. For many years, the impinger was the principal instrument used by industrial hygienists for industrial dust measurement. It would be almost 50 years before the impinger was to be replaced by other dust-sampling instruments.

The Vermont granite study also found that for some occupations, there was a 100 percent incidence of silicosis among workers with 15 or more years of exposure. Silicosis was found in some workers with only 2 years of service. Tuberculosis usually occurred after 20 years of service, and life expectancy was only approximately 6 months after the development of tuberculosis.

As a result of this study, the exposure standard recommended for silica dust was set at 5 MPPCF (million particles per cubic foot of air) in the size range of less than 10 microns. However, it took almost 10 years for the study recommendations to be implemented. A follow-up study in 1956 found that 15 percent of 1,390 granite shed workers had evidence of silicosis. Only one worker had evidence of silicosis among the 1,133 workers who began working in the granite sheds after engineering controls were begun in 1937.

4.1.2 The Uranium Mining Study

The study of uranium miners was initiated by the Public Health Service because it was expected that long exposure to radon would cause an elevated rate of lung cancer among these miners. This long-term study was started in 1950, when uranium mining was a relatively new segment of the U.S. mining industry. By 1963, an increased lung cancer incidence was evident among these uranium miners.

Early in the investigation, a tentative standard of 1 Working Level had been recommended. A Working Level (WL) was a combination of alpha-emitting products equivalent to 100 picocuries per liter of air. By 1969, data indicated that of the 6,000 men who had been uranium miners, an estimated 600 to 1,100 would die of lung cancer within the next 20 years because of radiation exposure on the job.

Most of the research in uranium was conducted by the Western Area Occupational Health Laboratory in Salt Lake City. The Public Health Service Division of Occupational Health recommended a standard of 0.3 WL for underground mines. This recommendation of 0.3 WL was later promulgated as a standard by the Bureau and the Department of Labor.

4.1.3 Coal Miner Health Studies

In 1926, the Governor of Pennsylvania and the United Mine Workers labor union asked the Public Health Service to undertake a study of coal miners' asthma. This investigation was conducted between 1928 and 1931 among the anthracite miners of eastern Pennsylvania by the Office of Industrial Hygiene and Sanitation of the Public Health Service. It was the first major epidemiologic study of chest diseases of coal miners in the United States.

Before the American study was completed, British investigators published findings that included a pathologic description of coal workers' lung disease that they called anthracosilicosis. The Public Health Service investigation determined that 23 percent of 2,711 miners examined had evidence of anthracosilicosis and suggested 25 MPPCF of coal mine dust as the recommended safe exposure. Studies by the Bureau in 1969 determined that 25 MPPCF of coal mine dust was about the equivalent of 2 mg/m^3, a value that was to become the legal standard in the Federal Coal Mine Health and Safety Act of 1969.

While recognizing a distinct disease entity in the anthracite coal mining population, many American investigators, prior to 1952, assumed that bituminous coal did not produce a disabling pneumoconiosis. There was a general belief that the pneumoconiosis reported by investigators was due primarily to the quartz content of the coal and was actually a type of silicosis.

Coal workers' pneumoconiosis (CWP) was recognized as a specific disease entity in about 1936 in Great Britain, when extensive studies were initiated by the Medical Research Council. The first studies of the health of bituminous coal miners in the United States began about 1941. The

Public Health Service conducted a study of bituminous coal miners in Utah in 1942, and found a low incidence of what was termed "anthracosilicosis" (3.2 percent). By 1949, as more studies were conducted, it was becoming apparent that CWP was a major problem among bituminous miners. Other limited confirmatory studies were conducted in 1957 and 1959 and provided further evidence of a disease problem.

In 1963, the Public Health Service began an extensive prevalence study that showed 10 percent of the active and 18 percent of the inactive miners had X-ray evidence of coal workers' pneumoconiosis. In 1967 the Public Health Service BOSH established the Appalachian Laboratory for Occupational Respiratory Diseases (ALFORD) on the campus of the West Virginia University School of Medicine in Morgantown, where most subsequent Public Health Service research on CWP would be conducted. Early research conducted at ALFORD included medical evaluations in coal miners to develop better clinical methods for diagnosing CWP and assessing the level of impairment caused by the disease. [Doyle 2009].

4.2 The USBM in the 1960s

By 1960, the mining research described in the Annual Progress Report of the Bureau of Mines fell into three principal areas [USBM 1960]: bituminous coal extraction, explosives and explosions, and health and safety. Under bituminous coal extraction, the report cited cooperation with industry in trials of a German coal planer for longwall mining, investigation of the use of coal augers in surface mines, high pressure water jets for coal extraction, and water infusion for methane degasification. Explosives and explosions research clearly was aimed at the safety of miners and other users of explosives. The most significant past accomplishment cited was the assessment and publicizing of the hazardous use of black powder in coal mining and the reduction in mine explosions resulting from its replacement by permissible explosives.

Other research areas cited in the report included:

- The practice of adding sodium chloride to coal mine explosives to reduce their ability to ignite methane.
- Investigation of flammability characteristics and explosion hazards of a wide variety of mineral, agricultural, and industrial dusts.
- Studies of coal dust transport, deposition, and explosibility, and methods for control of the coal dust explosion hazard.

In the Bureau's 1960 annual report [USBM, 1960], the following paragraph summarizes the accomplishments of the Bureau in the area of health and safety research and testing:

> These activities to promote safety and improve working environment throughout the mineral industries have been of a long-range, continuing nature. During previous years, notable accomplishments have been: (1) Development and widespread adoption of roof bolting for roof support in mines and tunnels, (2) development and application of new and improved methods of roof control (other than roof bolting), (3) determination of physiological action of a variety of gases and vapors, (4) determination of the occurrence, behavior, composition, physiological effects, and control of mineral dusts, (5)

development of performance requirements for industrial respiratory-protective equipment, (6) development of protective measures against mine explosions and fires, and (7) improvement of ventilation in mines and tunnels.

An internal Bureau report entitled "Research Facilities Program to the Year 2000" written in 1965 described the then research facilities existing in 1965 and contained a proposed plan for consolidation and expansion [USBM 1965]. In 1965, the Bureau owned and operated research facilities in the following locations:

- Tuscaloosa, AL
- Tucson, AZ
- Rifle, CO
- College Park, MD
- Minneapolis, MN
- Rolla, MO
- Boulder City, NV
- Reno, NV
- Grand Forks, ND
- Bartlesville, OK
- Albany, OR
- Bruceton, PA
- Pittsburgh, PA
- Schuylkill Haven, PA
- Norris, TN
- Salt Lake City, UT
- Morgantown, WV
- Laramie, WY

The Bureau also operated laboratories in leased space owned by others in the following locations:

- San Francisco, CA
- Berkeley, CA
- Anchorage, AL
- Denver, CO
- Mt. Hope, WV
- Spokane, WA
- Seattle, WA

The laboratories at Denver, Spokane, and San Francisco were in facilities administered by another federal agency, the General Services Administration.

Many Bureau research labs were located on or adjacent to the campuses of universities that included mining or metallurgy departments. The close proximity allowed for employment of faculty and students by the Bureau for its research and allowed Bureau staff to assist the faculty at the schools, establishing mutually beneficial arrangements to both the government and the universities.

The major activities of the Bureau in the mid-1960s were:

1. Research into the extraction, processing, and utilization of minerals, minerals fuels, and metals; and the safe use of explosives.
2. Development and conservation of minerals and fuel raw material resources.
3. Health and safety of persons engaged in the minerals industries.
4. Production and distribution of helium.
5. Special federal mineral industry programs, such as anthracite mine drainage and control of fires in inactive coal deposits.

Unlike many other federal research agencies, the Bureau's research programs were accomplished almost entirely in-house rather than by contracts or grants. In 1965, the Bureau had offices in 28 states and the District of Columbia.

4.3 USBM Research Laboratories in the 1960s

Most USBM research in the 1960s was focused on the efficient extraction, processing, and utilization of minerals and mineral fuels. Research primarily related to miner safety and health was performed at only a few labs, as described below.

4.3.1 Denver Mining Research Center, Denver, CO

This center performed basic and applied research on ground stabilization and control in mining, and the utilization of high-speed computers for the development of total mining systems. The Denver center maintained a close liaison with the Colorado School of Mines, located about six miles away.

4.3.2 Applied Physics Research Laboratory (Mining), College Park, MD

This lab specialized in theoretical laboratory and field studies on the behavior of rock formations under redistribution of stresses caused by mining, and also studies of stresses and rock breakage effects of blasting operations. This laboratory was closed in 1965 with the personnel and research transferred to Denver and Minneapolis.

4.3.3 Minneapolis Mining Research Center, Minneapolis, MN

This center performed mining research studies of rock penetration and fragmentation, including basic research in rock and mineral structures and their bonding systems.

4.3.4 Reno Mining Research Laboratory, Reno, NV

This lab undertook investigations into mining technology related to open-pit slope stability. The mining research and staff were transferred from Reno to Denver in 1966.

4.3.5 Explosives Research Center, Mining Research Center, Health and Safety Research Testing Center, Pittsburgh, PA

These three centers studied miner safety and health and were based primarily at the Bureau's Forbes Avenue facility in the Oakland section of Pittsburgh. Research areas included coal mining, control of mine water and methane gas, roof control in mines and tunnels, mining electrical equipment, health hazards and mine ventilation for their control, mine safety training, and other health and safety work. Several bills had already been introduced in Congress by 1965 to convey the Bureau land and buildings in the city of Pittsburgh to the Carnegie Institute of Technology. This transfer eventually took place in the 1980s.

4.3.6 Bruceton Laboratories, Bruceton, PA

Theoretical and applied studies on hazards associated with explosives, fires, and explosions were performed at the Bruceton Laboratories. The Bruceton facility was operated at this time as a joint entity with the Pittsburgh facility in Oakland. There were also extensive Bureau laboratories at Bruceton engaged in coal and energy-related research, and also a metallurgy laboratory performing iron and steel research. The 1965 Bureau plan advocated construction of new facilities at Bruceton in the expectation that the Forbes Avenue facility would eventually be transferred by Congress to the Carnegie Institute of Technology.

4.3.7 Spokane Mining Research Laboratory, Spokane, WA

The Spokane lab managed research to develop improved methods and systems for support of underground openings and the means for preventing rock bursts. Working relationships were maintained with nearby Gonzaga University, and with Washington State University and the University of Idaho, both about 80 miles away.

The 1965 report contained a proposal that would have eventually consolidated the Bureau's research to six locations: Washington, DC area, Minneapolis, MN, Bartlesville, OK, Albany, OR, Pittsburgh-Bruceton, PA, and Salt Lake City, UT. Interestingly, nearly 30 years later, a similar proposal would have consolidated all Bureau research to four centers at Minneapolis, Albany, Bruceton, and Salt Lake City. However, the Bureau plan for research facility consolidation in the 1990s was never carried out due to the closure of the U.S. Bureau of Mines in 1996.

4.4 Legislation and USBM Accomplishments of the 1960s

The Federal Metal and Nonmetallic Mine Safety Act of 1966 was the first federal statute directly regulating non-coal mines. This Act provided for the promulgation of standards, many of which were advisory, and for inspections and investigations of noncoal mines. However, it gave very little enforcement authority to the Bureau. The 1966 Act extended protection to miners at 18,000 noncoal mines [Kaas 2006].

The Bureau's annual progress report for 1969 provides a snapshot of the Bureau as it existed just before major changes spurred by major upcoming legislation, beginning with the Federal Coal Mine Health and Safety Act of 1969 [USBM 1969]. Research relevant to miner health and safety was reported under three separate areas: mining research, explosives and explosions research, and health and safety research. This separation of the research reporting reflected the organizational structure of the Bureau at that time.

In the Bureau report, some of the mining research accomplishments with direct safety implications were:

- Methods for predetermining the maximum stable angle of open-pit mine slopes.
- Microseismic methods to locate highly stressed areas and to determine the magnitude and location of rock bursts.
- An engineering manual delineating the application of rock mechanics principles for the design of underground openings.
- A methane control system that enabled more productive mining in a highly gassy mine using the longwall method.

Some of the explosives and explosions research accomplishments cited were:

- International accepted recommendations regarding the safe use of blasting agents.
- Guidelines for prevention of explosions from hundreds of flammable solids, liquids, and gases.
- Solving problems associated with utilization of combustion processes by defense agencies, NASA, and other government agencies, as well as industry. Such problems ranged from the flashback of gas appliance flames to the attitude control of the lunar excursion modules.
- Studies allowing for the safe use of ammonium nitrate-fuel oil blasting agents.

Finally, accomplishments reported under health and safety research included:

- Reduced incidence of mine disasters by Bureau development of rock dusting for prevention of dust explosions, foam for controlling mine fires, and more efficient mine ventilation systems.
- Use of fly ash for remote sealing to restore ventilation after mine explosions.
- Development of an instrument to warn of inadequate oxygen concentrations.
- Development of a methane monitor to warn miners or shut off power to a mining machine if dangerous levels of methane were detected.

In addition, the Bureau continued testing and approvals of permissible electrical equipment, diesel equipment, respiratory-protective equipment, and certification of explosives and blasting devices [USBM 1969].

5. Expansion and Change: 1969-1996

5.1 The Federal Coal Mine Health and Safety Act of 1969

On November 20, 1968, an explosion occurred at Consolidation Coal Co.'s No. 9 mine near Farmington, WV, resulting in the deaths of 78 miners (see Figure 32). At around the same time, medical studies were verifying that pneumoconiosis, the progressive respiratory disease caused by the inhalation of fine coal dust, was also a serious problem in coal miners. Pennsylvania had already adopted legislation to compensate coal miners with evidence of pneumoconiosis, but in some states coal workers' pneumoconiosis was not recognized as a compensable disease.

Figure 32. Smoke from the Farmington Mine disaster, Farmington, WV, 1968.

The Farmington disaster combined with the publicity about the prevalence of what miners came to know as "black lung" disease led to passage by Congress of the Federal Coal Mine Health and Safety Act of 1969 (1969 Coal Act). The 1969 Coal Act was more comprehensive and more stringent than any previous federal legislation governing the mining industry. The Act included surface as well as underground coal mines, required two annual inspections of every surface coal mine and four at every underground coal mine, and dramatically increased federal enforcement powers in coal mines. Safety standards for all coal mines were strengthened, and health standards were adopted for the first time. The Act included specific procedures for the development of improved mandatory health and safety standards, and provided compensation for miners who were totally and permanently disabled by "black lung."

The 1969 Coal Act was signed by President Nixon on December 30, 1969, and became law (Public Law 91-173). The new law provided for the assessment of civil and criminal penalties for violations of standards set under the Act. In 1970, there were about 2,400 active underground coal mines active in 22 states, and in March 1970 the Bureau had 327 mine inspectors, including 57 trainees. In response to the Act, by the end of the year the inspection force had increased to 564 persons, including 156 trainees. By mid-1971 the inspection force had increased to 1,000. Thus, the 1969 Coal Act caused the Bureau to triple its coal mine inspection force in just over a year. The Act also authorized the establishment of the Mining Health and Safety Academy at Beckley, WV, for the training of mine inspectors [CRS 1976].

The 1969 Coal Act was the first federal law to address both the health and safety of miners. The law required mine operators to take accurate samples of respirable dust in underground coal mines beginning June 30, 1970. Dust levels could be no higher than 3.0 mg of respirable dust per cubic meter, and the law required reductions to 2.0 mg per cubic meter within 3 years. The U.S. Bureau of Mines and the Bureau of Occupational Safety and Health (BOSH) of the Department of Health, Education and Welfare (HEW) jointly issued regulations specifying the procedures to be used for sampling for respirable dust in coal mines. BOSH and HEW were predecessors of the National Institute for Occupational Safety and Health and the Department of Health and Human Services (HHS). By the end of 1970, the Bureau had issued 714 violation notices to coal mine operators for excessive dust levels, and 1,589 notices for failure to initiate dust sampling programs.

The 1969 Coal Act also mandated a chest x-ray screening and surveillance program for coal workers' pneumoconiosis among underground miners, to be administered by HEW. The program was provided chest x-rays to all new underground coal miners and to all underground coal miners at specified intervals afterward. Miners with evidence of CWP were to be provided with enhanced dust sampling and/or transfer from high-dust to low-dust jobs in the mine.

The responsibility for this program was assigned to the ALFORD facility in Morgantown, WV. ALFORD scientists established an ongoing radiographic screening and surveillance program for all underground coal miners in the U.S. that continues to the current time. The 1969 Coal Act also mandated that epidemiological research be conducted to identify and define positive factors involved in occupational diseases of miners, provide information on the incidence and prevalence of pneumoconiosis and other respiratory ailments of miners, and improve mandatory health standards.

Over the next several decades, ALFORD scientists conducted a long-term, medical-environmental study of 5,000 coal miners in over 30 mines throughout the country. This study, the National Study of Coal Workers' Pneumoconiosis, was done in cooperation with the Bureau and was designed to obtain estimates of the number of miners with CWP, and to help develop a recommended standard for respirable coal mine dust [Doyle 2009].

The 1969 Coal Act also led the Bureau to reorganize in light of its increased responsibilities in mine health and safety. The Bureau was restructured under two deputy directors. A Deputy Director for Health and Safety with three Assistant Directors was responsible for the regulatory enforcement program of the Bureau. The three Assistant Directors oversaw coal mine health and safety, metal and nonmetal mine health and safety, and education and training.

The rest of the Bureau was placed under a second Deputy Director for Mineral Resources and Environmental Development, who had four assistant Directors. The responsibility for all mining health and safety research was placed under the Assistant Director for Mining. The Bureau's four mining research centers at Pittsburgh, Denver, Spokane, and Twin Cities were placed under the Assistant Director for Mining.

After the 1969 Coal Act, funding for coal mine health and safety research was greatly increased to $22.3 million ($119 million in current 2009 dollars) for fiscal year 1971. Health and safety research had previously been funded at levels between $1 million to $3 million during the 1960s. The increased health and safety research required increased staffing at the four mining centers in addition to the initiation of a large program of research contracts. In 1970, 44 contracts and grants were awarded totaling $7.9 million, and 27 of these were awarded to universities.

Despite the new law, the coal mining safety record for 1970 did not improve immediately. Two hundred fifty-five miners were killed in coal mines in 1970—52 more fatalities than in 1969.

Along with the 1969 Coal Act, in that same year the Bureau had contracted with the National Academy of Engineering to study ways to increase the probability of survival for miners in mine disasters. The report from the Academy recommended that the Bureau:

- Develop a portable self-contained escape breathing apparatus to provide oxygen for one hour.
- Investigate the feasibility of portable and central shelters with life support for 15 men for 15 days.
- Develop a communication system capable of transmitting coded messages to the surface.
- Develop a seismic system to locate a trapped miner who might be pounding on the mine roof.
- Acquire drilling rigs capable of drilling to 2,500 feet to probe for trapped miners and to provide for rescue [DOI 1970].

Finally, the 1969 Coal Act authorized the creation of two research advisory committees for mine health and safety. The Coal Mine Health Research Advisory Committee was created under the Department of HEW to advise the Secretary of HEW in the research needs related to coal miner health. In parallel, a Coal Mine Safety Research Advisory Committee was organized to advise the Secretary of the Interior on the research needs related to coal mine safety. Most members of both advisory committees were from the private sector, including academic researchers and members representing labor interests. The Bureau also had a representative on the Coal Mine Health Research Advisory Committee and HEW had an appointee on the Coal Mine Safety Research Advisory Committee.

5.2 Establishment of NIOSH

One year after the Federal Coal Mine Health and Safety Act of 1969, Congress extended broad safety and health protection to most other workers by passing the Occupational Safety and Health Act on December 29, 1970 (OSH Act). This Act created both the National Institute for Occupational Safety and Health (NIOSH) and the Occupational Safety and Health Administration (OSHA). OSHA, part of the U.S. Department of Labor, is responsible for developing and enforcing workplace safety and health regulations. NIOSH was authorized by the OSH Act to "conduct . . . research, experiments, and demonstrations relating to occupational safety and health" and to develop "innovative methods, techniques, and approaches for dealing with [those] problems" (Public Law 91-596, 84 STAT. 1590, 91st Congress, S.2193, December 29, 1970). NIOSH was authorized to conduct research, training, and education related to worker health and safety; perform on-site investigations to investigate hazards in the workplace; recommend occupational health and safety standards; and fund research by other agencies or private organizations. NIOSH was given the legislative responsibility to develop the research, based upon which it could then recommend occupational health and safety standards to OSHA and the Mine Safety and Health Administration (MSHA), although it did not have the authority to promulgate binding standards.

The immediate predecessor of NIOSH was the Bureau of Occupational Safety and Health, located in the Environmental Control Administration, an organization of the U.S. Public Health Service. Dr. Marcus M. Key, who was the Chief of the Bureau of Occupational Safety and Health, became the first Director of NIOSH. NIOSH was established for the specific purpose of conducting research that would lead to better prevention of work-related injury and illness. As an independent source of health-related research in the workplace, it was empowered to conduct industry-wide studies to identify illness and injury that might otherwise escape notice. NIOSH was to develop criteria for standards based on scientific data, and in turn, these criteria would be transmitted to the regulatory agencies, MSHA or OSHA, for consideration in their standard-setting process.

In the early 1970s, a substantial portion of the NIOSH budget was devoted to its responsibilities under the 1969 Coal Act. Subsequent to the establishment of NIOSH by the OSH Act of 1970, ALFORD moved into a new NIOSH laboratory building in Morgantown, WV, adjacent to the campus of West Virginia University. In 1976, ALFORD was reorganized and renamed as the Appalachian Laboratory for Occupational Safety and Health (ALOSH). The NIOSH Division of Respiratory Disease Studies (DRDS) was created as part of this reorganization, taking on many of the respiratory health programs arising from the 1969 Coal Act. These included the National Coal Workers' Health Surveillance Program, the National Coal Workers' Autopsy Study, and the National Study of Coal Workers' Pneumoconiosis.

Section 202 of the 1969 Coal Act created a statutory requirement that mine operators provide respirators for miners exposed to coal mine dust whenever the dust exceeded the established standard, and when they were exposed for short periods to inhalation hazards from gas, dust, fume, or mist. A respirator testing and approval regulation, 30 CFR 11, was prepared jointly by the Bureau and NIOSH and published March 25, 1972. This regulation consolidated the separate Bureau schedules into a single respirator testing and approval regulation, covering all types of respirators. The new regulation also imposed more stringent performance requirements on respirator manufacturers for approval of respirators.

Finally, a NIOSH-Bureau of Mines Memorandum of Understanding was signed in 1972, giving NIOSH sole responsibility for respirator testing, and the function of testing and certification of respirators was transferred to the NIOSH-ALFORD laboratory in Morgantown from the Bureau in Pittsburgh. Respirator testing, certification, and research became the responsibility of the newly formed NIOSH Division of Safety Research (DSR), a component of ALOSH, in 1976. As the result of a subsequent reorganization, this responsibility was moved to DRDS in 1997.

In 2001, almost 40 years after it had left for Morgantown, respirator testing, certification, and research returned to the Pittsburgh area when NIOSH established the National Personal Protective Technology Laboratory (NPPTL) at its Bruceton, PA, facility. NPPTL now has responsibility for respirator activities arising from the 1969 Coal Act and is an international leader in respirator testing, certification, and research.

5.3 Dismemberment of the USBM

On May 7, 1973, the Secretary of the Interior issued Secretarial Order No. 2953, creating a new agency within the Department of the Interior. Most of the health and safety enforcement functions were separated from the Bureau and placed in the new agency, which was named the Mining Enforcement and Safety Administration (MESA). Under the order, MESA became operative on July 16, 1973. The functions, staff, and facilities that had formerly been under the Deputy Director for Health and Safety in the Bureau were transferred to MESA. The functions transferred included mine health and safety inspection, technical support, assessment and compliance, and education and training. The new Mining Health and Safety Academy being constructed at Beckley, WV, was also eventually transferred to MESA. The functions of energy, metallurgical, and mining research, as well as the mineral supply information functions were retained by the Bureau, and the mining health and safety research program and staff remained with the Bureau. Approximately half of the facilities and staff of the Bureau were thus transferred to the new agency [DOI 1973].

The stated reason for the creation of MESA was to eliminate the possibility of a conflict of interest in the Bureau, because the Bureau mission to encourage the economic development of mineral resources might appear to conflict with the responsibility to enforce mining health and safety laws. The creation of the new agency left MESA free to devote its entire effort to the enforcement of mining health and safety laws.

By 1973, the safety record in coal mines was showing substantial improvement. The number of fatalities that year was 132, and the fatality rate was 0.45 per million man-hours worked—new record lows for these statistics. During 1973, the number of active underground coal mines also decreased from 1,663 to 1,425. American coal production in 1973 was 597 million tons, of which slightly more than half was from underground mines.

Expenditures by the Bureau for mining health and safety research in fiscal year 1973 were $28.4 million ($138 million in 2009 dollars). Some of the significant recent developments reported by the Bureau that year were:

- Degasification techniques for removal of methane from coalbeds ahead of mining.
- Machine-mounted systems for suppression of methane ignitions at the mining face.
- Explosion-proof bulkheads to preserve ventilation after explosions.
- Remote sealing of mine passageways to control fires.
- Canopies and cabs to protect equipment operators.
- Improved mine lighting and communications systems.
- Systems for locating trapped miners.
- A personal carbon monoxide alarm.
- Mine air monitoring systems.

Note that at this time a great deal of the research program was still directed at prevention and response to coal mine explosions and fires. The allocation of the $28.4 million of research funds in 1973 was as follows:

- Fire and explosion prevention—$5.0 million.
- Methane control—$2.6 million.
- Survival and rescue—$1.2 million.
- Industrial-type hazards—$5.3 million.
- Ground control—$5.9 million.
- Advanced mining systems—$3.7 million.
- Health—$4.7 million.

In 1973, the Bureau increased its efforts to transfer the results of health and safety research directly to the mining industry. In that year, the Bureau held five Open Industry Briefings in mining areas to communicate research results, as well as four seminars on specific safety technologies. The Bureau also initiated a new publication series called *Technology News*. These were concise descriptions of individual technology developments printed on a single sheet of paper. They were intended to provide timely information of the latest technologies in the mining industry.

The Annual Progress Report of the Bureau of Mines for Fiscal Year 1973 [USBM, 1973] appears to have been the last annual fiscal year report by the Bureau, ending a series that had begun with the establishment of the Bureau in 1910. This report mentioned several other publications that also reported on the Bureau's work, and these publications appeared to replace the fiscal year annual report after 1973. These other publications included:

- "Mining and Minerals Policy 1973," an annual report under the Minerals Policy Act of 1970.
- "Research 72," a new annual report series on all Bureau research programs.
- "Administration of the Federal Coal Mine Health and Safety Act," an annual report to Congress on progress under the 1969 Coal Mine Act, and also a similar report to Congress for the Federal Metal and Nonmetallic Mine Safety Act of 1966.

The 1973 report continued to classify safety-related research under three separate sections: Mining Research, Explosives and Explosions Research, and Health and Safety Research. The Mining Research program of the Bureau had been conducted for many years prior to the 1969 Coal Act. Most of the work classified then as mining research was in the fields of rock mechanics, blasting theory and practice, mine waste disposal, ground support, drilling, rock fragmentation, materials handling, methane control, and acid mine water control. Note that most of this Bureau mining research was related to miner safety, even though it was often justified as also increasing mining efficiency or productivity.

The 1973 report cited the previous accomplishments for Mining Research as follows:

> Pit slope stability studies which established the maximum stable angle of open-pit mine slopes, blasting studies which established the well-accepted theory of reflection breakage,

vibration studies which became the accepted reference in the blasting field and are the basis for many state blasting codes, mine waste disposal studies which established the engineering basis for waste embankment regulation under the Health and Safety Act, the microseismic work which locates incipient rock bursts, the distressing studies which may someday prevent rock bursts, a low-cost acid mine water treatment technique, model studies which led to new concepts in the design of mine pillars and mine pillar reinforcement, and in-mine studies of pillar strength which will permit smaller mine pillars and greater productivity without sacrificing safety [USBM 1973].

The work done under Explosives and Explosions Research was intended to improve the safety and efficiency of explosives used in mining, to provide maximum safety in their production and handling, and to prevent accidental fires and explosions. Some of the functions of this work included:

- Administering federal regulations for certification of permissible explosives.
- Conducting basic and applied research to understand explosion processes.
- Developing practical methods to prevent and control mine fires.
- Providing consultation on explosions and on safe use of explosives to other government agencies and industry.

Some of the previous accomplishments cited for this Bureau program were:

- Development of the new permissible explosives and testing procedures for these explosives that had greatly improved the safety of blasting with explosives in mines.
- Recommendations for storage, transport, and utilization of blasting agents and hazardous chemicals that had received national and international recognition, including a recent study for the U.S. Coast Guard of hazards from spills of liquefied natural gas.
- Development of a theoretical model of the detonation mechanism that had helped make the Bureau a world authority in the field of explosives research.

Some of the previous accomplishments cited for Health and Safety Research were:

- Studies on dust collectors, wetting agents, and foams for respirable dust control.
- Use of vertical drainage holes for methane control.
- Techniques for drilling and packing horizontal holes in coalbeds to help measure and control methane.
- Technology for design and evaluation of roof control plans for mines using roof bolts for support.
- Lighting studies to provide the basis for proposed mandatory illumination standards.
- A number of devices developed for improved safety in underground coal mines.

The funds appropriated to the Bureau by Congress for fiscal year 1973 totaled $157 million. This funding can be broken out into the following areas:

- Health and safety enforcement—$64 million
- Energy-related research—$20 million

- Energy-related minerals information—$7 million
- Non-fuel minerals information—$10 million
- Metallurgy research—$16 million
- Mining, explosives, health and safety research—$38 million
- General administration—$2 million

The health and safety enforcement area was split out to MSHA in 1973, and the two energy-related functions became part of the Department of Energy (DOE) by 1977. Thus, $91 million, or approximately 58 percent of the Bureau's fiscal year 1973 budget, was in functional areas that were to be split off from the Bureau in the period 1973-77.

5.3.1 Transfer of USBM Research Labs to ERDA

In 1975, the Bartlesville Energy Research Center, the Morgantown Energy Research Center, and the Pittsburgh Energy Research Center, as well as a few smaller energy research labs, were transferred from the Bureau to the new U.S. Energy Research and Development Administration (ERDA). ERDA was created as part of the Energy Reorganization Act of 1974, passed on October 11, 1974, in the wake of the 1973 oil crisis.

ERDA was formed mostly from the split of the Atomic Energy Commission (AEC), and it was given the research functions of the AEC not assumed by the Nuclear Regulatory Commission. ERDA was activated on January 19, 1975. In 1977, ERDA was combined with the Federal Energy Administration to form the United States DOE. Under DOE, the three former Bureau sites became the Bartlesville Energy Technology Center (BETC), Morgantown Energy Technology Center (METC), and Pittsburgh Energy Technology Center (PETC). The centers' responsibilities included in-house research in coal, oil, and gas technologies as well as management of millions of dollars in contracts for research and development conducted by universities, industry, and other research institutions. These centers formed the basis of the current National Energy Technology Laboratory of DOE.

5.3.2 Coal Advanced Mining Technology Program

Following the oil embargo crisis of 1973-74, the Bureau began a large new program of coal mining research and development (R&D) to allow for a rapid expansion of the coal industry if needed to respond to any future energy crisis. This was the Advanced Mining Technology Program, and it rapidly expanded with a budget for fiscal year 1976 of $44.5 million ($169 million in 2009 dollars) as compared to $34.9 million for all mining health and safety research in that year. Some of this expanded funding was used for the construction of new mining research buildings and facilities at Bruceton, including the collection of buildings and equipment that was called the Mining Equipment Test Facility (METF). Much of this expanded research was planned to be performed at a new Bureau mining research center to be established in Carbondale, IL.

When the Department of Energy was established in 1977, the Coal Advanced Mining Technology Program was transferred to DOE along with the newly established research center at Carbondale, as well as the staff and facilities of the program that were located at Bruceton. The

Coal Advanced Mining Technology Program was reduced in budget by DOE over the next few years, eventually transferred back to the Bureau along with the remaining staff and facilities in the early 1980s. By that time, the program had been reduced to only a small fraction of its former size when it had left the Bureau.

The Secretary of the Interior established the Office of Surface Mining, Reclamation, and Enforcement (OSM) in 1977. This new agency was responsible for regulation of coal mining under the Surface Mining Control and Reclamation Act of 1977 (SMCRA). OSM was given three main functions: (1) regulating active mines, (2) reclaiming lands damaged by surface mining and abandoned mines, and (3) providing resources for technical assistance, training, and technology development. The Bureau continued to conduct mining research, including the new Minerals Environmental Technology Research Program that it established in 1979.

5.4 Federal Mine Safety and Health Act of 1977

The Federal Mine Safety and Health Act of 1977 (the 1977 Mine Act) (Public Law 95-164) amended the Coal Mine Safety and Health Act of 1969. Enacted November 9, 1977, it took effect 120 days later. The 1977 Mine Act extended most of the provisions of the 1969 Coal Act to all other mines and also increased the authorized funding for health and safety research. The 1977 Mine Act amended the 1969 Coal Act in a number of significant ways, principally by consolidating all federal health and safety regulations of the mining industry—coal as well as non-coal mining—under a single law.

The 1977 Mine Act also strengthened and expanded the rights of miners, enhancing the protection of miners from retaliation for exercising their rights. The 1977 Mine Act transferred MESA and its functions from the Department of the Interior to the Department of Labor, and changed the name of the agency to the Mine Safety and Health Administration (MSHA). Additionally, the 1977 Mine Act established the independent Federal Mine Safety and Health Review Commission to provide for independent review of the majority of MSHA's enforcement actions (MSHA website, 03/10/09).

The organizational effects of the 1977 Mine Act on MESA were to transfer the entire agency to the Department of Labor, as well as to change its name to MSHA to parallel its sister agency in that department, the Occupational Safety and Health Administration (OSHA). MSHA was given regulatory authorities for noncoal mines similar to those it had already possessed for coal mines. MSHA and OSHA negotiated an agreement that defined their areas of responsibility in metal and nonmetal mines and mills. The agreement specified that MSHA had responsibility for regulation on mining property until such stage in the process that chemical changes were made in the ore being produced; OSHA had responsibility for safety and health regulation for any later stages in mineral processing.

The Federal Mine Safety and Health Amendments Act of 1977 authorized NIOSH to:

- Develop recommendations for mine health standards for the Mine Safety and Health Administration;

- Administer a medical surveillance program for coal miners, including chest X-rays to detect pneumoconiosis in coal miners;
- Conduct on-site investigations in mines similar to those authorized for general industry under the OSH Act;
- Test and certify personal protective equipment and hazard-measurement instruments.

In relation to health research, the 1977 Mine Act also empowered NIOSH with much greater responsibility within the mining industry, since its research mandate was extended to include the health of all miners, not just coal miners. Previous NIOSH research had been devoted to the health of noncoal miners only under the authority of the Public Health Service Act (Public Law 78-410). The Bureau had previously done mining health and safety research in noncoal mines under authority of its Organic Act of 1913 and also the Federal Metal and Nonmetallic Mine Safety Act of 1966 (Public Law 89-577).

The Federal Mine Safety and Health Amendments Act of 1977 (Public Law 95-164) had given responsibility for mine health research to NIOSH, through the Secretary of the Department of Health, Education, and Welfare, and responsibility for mine safety research to the Bureau, through the Department of the Interior. The Bureau also conducted research on mining technology aimed at protecting the health of miners.

A Memorandum of Understanding was signed by the acting directors of NIOSH and the Bureau in the summer of 1978. This agreement set guidelines for division of research responsibilities between the two agencies, specifying how the agencies would interact to coordinate their research programs in mining health and safety. In general, the agreement specified that the Bureau would conduct mining research designed to improve safety and to improve the environmental conditions in mines. The Bureau's health-related research was to study technology to reduce the health hazards of miners, such as improving mining methods, equipment, and work practices. NIOSH would conduct research to prevent mining-related occupational diseases through examination of miners, evaluation of their exposures to health hazards, and development of improved occupational health methodology. Under this agreement, NIOSH and the Bureau coordinated their research activities, especially in the area of miners' health, until the absorption of the Bureau's mining health and safety research into NIOSH in 1996.

The 1969 Coal Act (Sec. 102) had authorized the Secretary of the Interior to appoint an advisory committee on coal mine safety research and the Secretary of HEW to appoint an advisory committee on coal mine health research. During 1970, the Secretary of HEW established and appointed members to the Secretary's Coal Mine Health Research Advisory Council. The Council's first meeting was held on November 10-11, 1970, in Washington, D.C. The Council made recommendations to the Secretary on matters relating to coal mine health research undertaken under the 1969 Coal Act including intramural research, research grants, and contracts.

The Secretary of Interior also established an advisory committee on coal mine safety research, which operated from 1970 to 1977. The 1977 Mine Act authorized committees on both mining safety research and health research. The Department of the Interior did not establish a new committee of mining safety research under the 1977 Mine Act, allowing the coal mine safety research advisory committee to expire.

The Secretary of HEW established the Mine Health Research Advisory Committee (MHRAC), which operated from 1977 to 1996. After the closure of the U.S. Bureau of Mines in 1996 and the assumption by NIOSH of both the safety and health research functions, MHRAC was reestablished in 1997 as the Mine Safety and Health Research Advisory Committee (MSHRAC), and MSHRAC continues to advise NIOSH on all research under the 1977 Mine Act.

5.5 USMB Research Centers after 1979 Reorganization

Following the 1969 Coal Act, the USBM enjoyed a large increase in its budget for coal mining health and safety research. Most mining health and safety research was completed at four of the Bureau's research locations in Pittsburgh, PA, Minneapolis, MN, Denver, CO, and Spokane, WA. Brief histories of these four research facilities follow, which were re-named following a 1979 as follows: Pittsburgh Research Center, Twin Cities Research Center, Denver Research Center, and Spokane Research Center.

5.5.1 Pittsburgh Research Center

The Pittsburgh Research Center of the Bureau originated with the beginnings of mine safety research at the Government Arsenal in the Lawrenceville section of Pittsburgh in 1908. This work began under the Technological Branch of the US Geological Survey and continued with the transfer of this branch to become the basis of the new U.S. Bureau of Mines in 1910. Initially, the site was called the Pittsburgh Mining Experiment Station. In 1917, the work was moved to the newly constructed Central Experiment Station at 4800 Forbes Avenue in the Oakland section of Pittsburgh, near the Carnegie Institute of Technology and the University of Pittsburgh. Explosives and coal mine explosions research was completed at the Bureau's location at Bruceton (see Figure 33), which included an experimental mine, and the Bruceton site was called the Explosives Experiment Station.

Figure 33. Aerial view of the Bruceton facility in PA, 1988.

Over the years, the federal presence at Bruceton expanded as the government purchased 72 acres around the experimental mine in 1924 and purchased 146 additional acres at the site in the 1940s and 1950s. The large acreage was necessary because of the dangers of storing explosives on site and the explosion tests conducted at the experimental mine. By 1970, Bureau research continued to be divided between the locations at Bruceton and at Forbes Avenue. During the early 1970s, research was consolidated at Bruceton, and the Forbes Avenue facility was used primarily by the inspectorate part of the Bureau, which eventually became part of MESA and then MSHA.

The Lake Lynn Laboratory facility was opened by the Bureau in 1982 (see Figure 34). Lake Lynn Laboratory is a 400-acre facility near Fairchance, PA, about 50 miles south of Pittsburgh on the Pennsylvania-West Virginia state boundary. The laboratory was constructed by driving new underground entries adjacent to the workings of an abandoned limestone mine. The new entries simulate the configurations found in modern underground coal mines.

Figure 34. Main portal for the Lake Lynn Laboratory near Fairchance, PA.

Lake Lynn Laboratory's primary purpose was as a new experimental mine for mine explosion testing. Due to the proximity of dwellings built as suburban Pittsburgh expanded to the Bruceton area, the experimental mine at Bruceton could no longer be used for testing large explosions. The Lake Lynn facility was also used for many other types of mining research. The surface facilities at Lake Lynn are in an isolated environment, which allows for large-scale testing in an environmentally controlled manner.

5.5.2 Twin Cities Research Center

The Twin Cities Research Center began with Congress authorizing a Bureau metallurgical research station at the University of Minnesota at Minneapolis in 1915. In 1920, this location was called the North Central Experiment Station. Mining research began shortly after the end of World War II in a new Minneapolis branch of the Bureau's Mining Division in downtown Minneapolis. The Bureau acquired land in 1951 at the Fort Snelling site in Minneapolis on the banks of the Mississippi River near the airport. New buildings were constructed at the Fort Snelling site and the main building was opened in 1959, housing both metallurgical and mining research centers of the Bureau. In the 1979 reorganization, the Metallurgical and Mining centers were merged and the combined laboratory was named the Twin Cities Research Center.

The early focus of mining research at Twin Cities Research Center was on novel methods of rock fragmentation. By the mid-1980s, Twin Cities Research Center research included coal and rock fragmentation, advanced fragmentation techniques, rock physics, drilling technology, blasting, in-situ mining and leaching, mine hydrology, waste water technology, vibrations from surface blasting, diesel emissions, surface mine equipment safety, fire protection, and respirable dust generation.

5.5.3 Spokane Research Center

In 1951, the Bureau's mining division in Albany, OR, was moved to Spokane, WA, to locate the work nearer to the deep mines of Idaho and Montana. This mining division was renamed the Spokane Research Laboratory in 1963. Early mining research in Spokane focused on mining extraction methods and equipment and was intended to improve both the safety and efficiency of mining. The research emphasized regional problems of the deep lead and silver mines in the Coeur d'Alene mining district of northern Idaho, which is near Spokane. These problems existed primarily in ground support, rock burst control, and mine waste disposal. The site was renamed the Spokane Mining Research Center in 1971 and in 1979 became the Spokane Research Center. The center was located on a 3.4-acre site just north of downtown Spokane (see Figure 35). Bureau research performed at Spokane in the 1970s and 1980s included ground control, radon control, mine productivity, mining with backfill, reclamation, and waste disposal technology.

Figure 35. Spokane Research Center, Spokane, WA, circa 1985.

5.5.4 Denver Research Center

The Bureau opened a Denver field office of its Safety Division in 1910. This office assisted in the training of miners in accident prevention, mine rescue, and first aid. In its first decade, the Bureau also opened a rare metals experiment station in Denver with a primary focus on methods for efficient processing of radium, which at the time was considered to be a promising substance for treating cancer patients.

In 1916, the research offices were moved to the campus of the Colorado School of Mines at Golden, CO, near Denver. Mining research at Denver began with the establishment of an oil shale mining branch at Denver following passage of the Synthetic Liquid Fuels Act of 1944. The Bureau also constructed an experimental oil shale mine and demonstration plant at Rifle, CO, at this time.

In 1949, Bureau research near Denver was consolidated in leased buildings at the Denver Federal Center in Lakewood, CO, a western suburb of Denver. In the 1960s, the Denver work was expanded with the transfer of the Bureau's Applied Physics Laboratory from College Park, MD, to Denver, and the similar transfer to Denver of the mining research laboratory from Reno, NV. In the 1979 reorganization, the name of this Bureau facility was changed to the Denver Research Center. The facilities of the Denver Research Center also included the Twilight Mine, a small leased experimental uranium mine in western Colorado.

Much of the research at Denver after 1970 focused on solving ground control problems in both coal and metal/nonmetal mines, with the work performed by researchers with expertise in geophysics and rock mechanics. The Bureau also did extensive research at Denver on methods for measurement and control of radon progeny—a cause of lung cancer and a significant problem in underground uranium mines. The Bureau's experimental uranium mine was extensively used by Bureau researchers from Denver and Spokane who were working on the radon control problem.

5.5.5 Mineral Institutes

The Mineral Institutes program of the Bureau was authorized by Congress in 1984 under Public Law 98-409, the State Mining and Mineral Resources Research Institute Program Act. This law authorized Bureau grants to universities with substantial graduate programs in the mineral sciences and engineering. Awarded universities were designated as State Mining and Mineral Resources Research Institutes (Mineral Institutes). Eventually, 31 universities were designated as Mineral Institutes and received annual allotment grants from the Bureau for research and educational programs.

Allotment grant funds from the Bureau were used to support fellowships and scholarships, to acquire scientific equipment, and to support small mineral-related research projects. The annual grants were to be matched with two local dollars for each dollar of federal grant aid. The designated Mineral Institutes were also eligible to compete for research support through the associated Generic Mineral Technology Center (GMTC) program.

Six GMTCs were established to conduct research in several broad engineering areas of significance to the minerals industry. Each GMTC was headquartered at one university and had several affiliate universities designated as Mineral Institutes. Two of the Generic Centers were involved with research on miner health and safety. These were the Generic Center for Respirable Dust headed jointly by Penn State University and West Virginia University, and the

Generic Center for Mine Systems Design and Ground Control, led by Virginia Tech. Funding for the Mineral Institutes program and the GMTCs continued until the closure of the Bureau in 1996.

5.6 USBM Accomplishments in Mining Research

Many USBM research and accomplishments in mining safety occurred in the last 25 years of its history, between the passage of the Federal Coal Mine Health and Safety Act of 1969 and up until the Bureau was terminated by Congressional action in 1996. During those years, mining health and safety research was divided into the following subprogram areas:

- Health research: respirable dust, noise, radiation, and diesel emission control.
- Safety research: ground control, human factors, industrial safety, and disaster prevention.

Three of the largest subprograms were respirable dust, ground control, and disaster prevention. A few of the most significant accomplishments of Bureau research after the 1969 Coal Act will be described next in more detail [USBM 1993].

5.6.1 Respirable Dust

The Bureau helped develop several devices for measurement of airborne dusts in mines. One of these, the RAM-1 light-scattering dust monitor, was developed under contract by GCA Corporation for the Bureau in 1978. This instrument was used by Bureau engineers and others to evaluate the effectiveness of new dust control techniques in mines, offering the major improvement of continuous readings of dust concentrations in real-time. GCA later designed a smaller passive dust monitor for the Bureau, called the MINIRAM, which became commercially available and has been used extensively. The Bureau and NIOSH also jointly funded a contract with GCA for a device called the Fibrous Aerosol Monitor to measure the airborne concentration of fibrous aerosols such as asbestos.

In the 1970s, the Bureau did major research on dust scrubber systems for coal mines, helping to develop a flooded-bed scrubber that was more than 95 percent efficient in removing respirable dust from mine air. Flooded-bed scrubbers are now used on most continuous mining machines in underground coal mines in the U.S.

Much of the Bureau's dust control research in the 1970s and 1980s was aimed at controlling dust on longwall coal mine faces, as this was the type of mining operation with the greatest difficulty in maintaining compliance with the regulatory dust standards. One method developed for longwall shearers was the "shearer-clearer" system, which oriented water sprays on the shearer mining machine to entrain air and hold the dust-laden air near the coal face and away from the machine operators, who could thus operate the machine in relatively clean intake air (see Figure 36). All longwall shearers in the U.S. now use such spray systems.

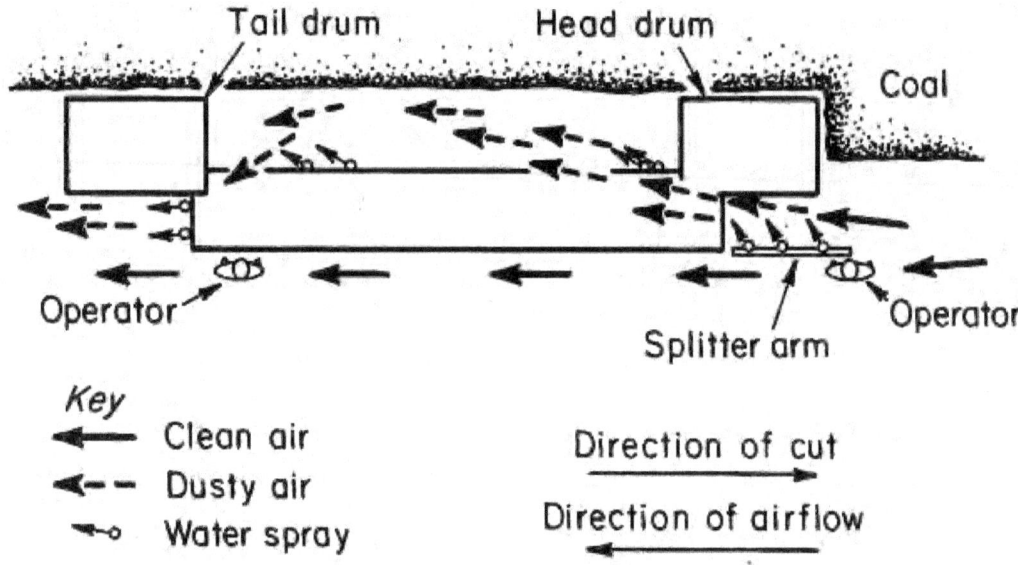

Figure 36. Air currents when using the shearer-clearer system.

The Bureau's in-house and contract respirable dust research program was assisted by the GMTC for Respirable Dust. This group of universities led by the Penn State University and West Virginia University conducted government-funded research on the measurement and control of respirable dust and other hazardous aerosols in mines. The GMTC for Respirable Dust research also included biomedical studies that helped increase understanding of the causes and mechanisms of coal workers' pneumoconiosis. This critical work was closely coordinated with that of the Bureau and NIOSH. The GMTC for Respirable Dust not only produced a great deal of important research results, but also helped train dozens of graduate students for work in the mining industry and elsewhere.

5.6.2 Ground Control

One of the largest subprograms of the Bureau in this period was ground control research to prevent injuries from falls of roof and rib in underground mines, which are generally a leading cause of mining fatalities. Research on developing instrumentation and methods to detect hazardous ground conditions included work on borehole extensometers, continuous roof displacement recorders, and mechanical and ultrasonic convergence meters. These devices measured roof to floor convergence and roof strata separation that could indicate imminent roof failure. Geophysical techniques, including microseismic monitoring, were studied that could be used to help warn of impending roof failure. Microseismic activity or acoustic emissions are low-level noises that emanate from stressed rock in underground mines. This approach was discovered and studied by two Bureau researchers, Leonard Obert and Wilbur Duvall, in the 1930s.

After the 1969 Coal Act, the Bureau continued research to improve the roof bolting process that had become the principal roof support method in American underground mines. Bureau research helped establish industry practices for anchorage and load capacity for various roof bolting systems. Investigations were made into new roof control concepts including a pumpable roof bolt, a yielding roof bolt, a flexible helical rock bolt, and various bolt grouts. These studies provided valuable information to the mining industry to improve roof support effectiveness. Other roof support systems the Bureau investigated included a mobile shield, automated temporary roof support systems, continuous miner-based automated roof support systems, steel-fiber reinforced concrete cribs, and improved arch canopies.

Retreat room-and-pillar mining, where a series of pillars is developed as a section is advanced with pillars later extracted upon retreat, is a common mining method used to extract coal from underground mines. Although longwall mining has overtaken room-and-pillar mining in popularity, pillar recovery still accounts for about 10 percent of U.S. underground coal mining. In the past, pillar recovery had been associated with a disproportionate rate of roof fall fatalities. During 1980-1997, nearly 25 percent of all roof fall fatalities occurred on pillar recovery sections. The main danger was premature caving of the roof at the face, where mining crews were working to install wood posts as the coal pillars were extracted.

Research by Bureau engineers in the 1980s resulted in the development of mobile roof supports (MRSs). These remotely controlled, hydraulically operated systems support the mine roof during pillar extraction and replace the use of wooden posts. The successful demonstration of MRS technology led to commercialization by the mining manufacturer, J. H. Fletcher and Co. However, the MRS did not provide critical feedback to workers of impending ground failure as do the wood posts, which begin cracking and "popping" under load. Therefore, MSHA asked that guidelines be created for safe use of the MRS and that a warning system be developed to alert miners to impending roof caving.

In response, NIOSH later developed and field tested a warning system for the MRS that would alert miners to unstable roof and pillar conditions. The warning system provides a visual indication of current loading as well as changes in loading on the MRS. An increase in load rate tells miners that the pillars are yielding and failing, as the loads once carried by the pillars are transferred to the MRS. This condition often triggers failure of the mine roof.

The use of MRSs in U.S. retreat mines has grown dramatically in recent years. In 1988, only four MRSs were in use. The number has grown to about 270 currently in operation at 34 U.S. coal mines. The successful demonstration of the load-warning system resulted in the incorporation of this system into the design of the Fletcher MRS.

During the 1970s, the Bureau constructed a mine roof simulator at the Bruceton laboratory for the testing of longwall shields and other types of roof support devices (see Figure 37). The Mine Roof Simulator is a unique biaxial load frame that provides controlled load conditions of a strength sufficient to simulate the roof loading conditions of underground mines. The Bureau did extensive research on longwall shields and provided valuable basic information used by manufacturers to design safer shields for supporting the roof of longwall faces. The Mine Roof Simulator has been upgraded and is used extensively in the NIOSH mining research program.

Figure 37. Testing of the mine roof simulator.

Bureau research in ground control was supplemented by the Generic Mineral Technology Center for Mines Systems Design and Ground Control. This was a Bureau-funded group of universities led by Virginia Tech that researched a wide range of topics to improve mine design safety and efficiency, with particular emphasis on ground control systems.

5.6.3 Disaster Prevention

The Bureau's disaster prevention subprogram included the areas of methane control, fires and explosions, life support, explosives, and mine rescue.

Methane explosions have been the major cause of multiple fatality incidents in underground coal mines. In the first decade of the 20th century, as many as 500 miners were killed each year in coal mine explosions. This death rate was gradually reduced by improvements in coal mining technology and practices, including ventilation, removal of ignition sources, use of permissible explosives and explosion-proof electrical equipment, improved methane measurement devices, and control of float coal dust that could amplify methane explosions. Many of these improvements were outcomes of research by the Bureau.

After the 1969 Coal Act, the Bureau did extensive research on methods of methane drainage to help maintain safe methane levels in underground coal mines. These methods included drilling horizontal holes into unmined coal in advance of mining, and vertical gob gas vent holes drilled from the surface into strata just above the coal to be mined. Drilling horizontal holes can provide immediate relief from methane problems in existing mines and are particularly useful for longwall mines. Drilling vertical gob gas vent holes into strata above longwall panels can also help remove methane that otherwise could enter the working areas of coal mines.

Bureau research on methane drainage in the 1970s and early 1980s helped lead the widespread use of this technology today. This technology has helped make mine gas explosions relatively rare, and the methane drained from the mines has been a welcome addition to U.S. natural gas supplies.

The Bureau also developed the direct method test to measure the gas content of coal samples. This method allows mining companies to determine the gassiness of their coal reserves during exploratory drilling—a direct aid to ventilation system and methane drainage planning before mine development begins. The portable modified direct method apparatus developed by the Bureau won an R&D 100 award in 1997.

In 1972, an underground mine fire at the Sunshine Mine in Kellogg, ID, resulted in the deaths of 91 miners. This disaster was one of the factors that led to the passage of the 1977 Mine Act. It also led to an immediate expansion in Bureau research on prevention and control of underground mine fires. The Bureau began new research on test procedures for the evaluation of fire resistance of materials used underground, fire detection instrumentation, computer models to predict the spread of fires and toxic gases, and techniques for sealing and extinguishing underground fires.

In addition, the Bureau did research on the flammability of mine ventilation tubing that helped lead to the adoption of a new federal standard test procedure. A new laboratory scale test procedure was also developed to test the fire resistance of conveyor belt materials. The Bureau-developed mine fire simulation computer program, MFIRE, was used to model the spread of mine fires. This software is used to predict the spread of combustion products through underground mine passages. It provides great value in the design of fire detection systems and for mine escape and firefighting strategies.

Finally, Bureau research in the 1970s and 1980s helped to develop a self-contained self-rescuer (SCSR), allowing miners to escape from underground mines when the mine atmosphere contained toxic gases or insufficient oxygen. Previously, miners wore filter self-rescuers that only protected against low levels of carbon monoxide, but not against other toxic gases or oxygen deficiency. Combined Bureau and the private sector research also led to the development of several SCSRs that could provide a sixty-minute supply of oxygen to help miners escape in an emergency. Federal regulations required that SCSRs be available in the face area of underground coal mines, and they began to be deployed in mines in October 1982.

5.6.4 Illumination

Before the 1969 Coal Act, the only illumination in underground coal mines was often the cap lamp worn by the miners. The electric cap lamp was a great safety improvement over the open flame lamps used early in the century; however, it provided only a spot of light directly in front of the miner, and was little help for peripheral vision. The 1969 Act specifically called for improvements in the illumination of underground coal mines. Bureau research determined that 0.06 footlamberts was the minimum light level needed for a safe workplace. The Bureau also developed several new illumination systems and demonstrated them in operating coal mines. These systems were rugged enough to survive in a mining environment and also explosion-proof.

5.7 R&D 100 awards and the USBM

One measure of the accomplishments and impact of a research program is recognition of its products by the R&D 100 awards program of *R&D Magazine*. These awards are presented annually for the 100 most technologically significant new products. In total, the USBM received 33 R&D 100 Awards before being closed in 1996, and ranked 13th in the world in the number of these prestigious awards [Kirk 1996]. The USBM health and safety research program received 16 of these awards, as follows:

- Fibrous Aerosol Monitor (1978)
- Reusable Ventilation Control Stopping for Underground Mines (1978)
- Trapped Miner Electromagnetic Transmitter (1979)
- Self-Rescue Breathing Apparatus (1980)
- Palmes Passive NOx Samplers (1980)
- Cement-Grouted Roof Bolt (1980)
- Flexible Roof Drill (1981)
- Geodynamic Accumulated Strain Sensor (1981)
- Sheathed Explosive Charge (1982)
- Wireless Real-Time Survey Tool (1982)
- Downhole Replaceable Diamond Core Drill Bit (1982)
- Rock Dust Meter (1985)
- Ignition Suppression System (1985)
- Personal Toxic Gas Alarm—"PERTGAL" (1989)
- SR-100 Ultra-Lightweight Rebreather (1991)
- "Canary" Wireless Mine Messenger Communication and Emergency Warning System (1993)

In addition, after the mining health and safety research program was transferred from the Bureau to NIOSH, several more awards were received:

- Portable Modified Direct Method Apparatus (1997)
- Personal Dust Monitor—PDM (2004)
- Coal Dust Explosibility Meter—CDEM100 (2006)

5.8 Decline and Dissolution of the USBM

The existence of the U.S. Bureau of Mines as a separate federal agency had been threatened since its dismemberment began in 1973 with the separation of its enforcement function and the creation of the Mine Safety and Enforcement Agency (MESA) that later became MSHA. The Bureau lost more of its other functions in 1974 to the newly established Energy Research and Development Agency (ERDA), and in 1977 to the new Department of Energy (DOE).

One of the bills passed by Congress in 1976 to create a Department of Energy would have entirely absorbed the functions of the Bureau into that new department, abolishing the Bureau as it then existed. That bill was vetoed by President Ford. The law establishing DOE in 1977, signed by President Carter, retained the Bureau as a separate agency within the Department of the Interior. However, it did move additional functions from the Bureau to DOE, including research on improving mining efficiency. Some of the Bureau staff and facilities devoted to mining efficiency research were transferred to DOE in 1977. Most of these were transferred back to the Bureau in the early 1980s.

By 1993, the Bureau's future did not look promising. Director TS Ary left the Bureau at the beginning of the Clinton Administration in January 1993. The continued existence of the Bureau as an independent agency appeared uncertain, as no new Director was named for one and a half years. Finally, President Clinton nominated Rhea Graham for the position in August 1994.

During this time, the Deputy Director, Dr. Herman Enzer, was the acting Director of the Bureau. He initiated a review of Bureau programs that resulted in a report, "Reinventing the USBM" in 1994. This report recommended major changes both in the organization and programs of the Bureau, particularly in research. In fact, the report seemed to indicate a shift of emphasis toward environmental research and away from research devoted to miner health and safety. In discussing health and safety technology, the report cited the following: "Significant problems that existed 20 years ago have largely been solved or at least mitigated; research on these should be reduced or eliminated" [USBM 1994].

Another report recommendation was to consolidate the mining health and safety research of the Bureau at one "center of excellence" to be located at Pittsburgh. The Denver and Spokane labs would have been made satellite centers that would continue their existing research, but were considered likely to be reduced or eliminated in any future budget reductions. The Twin Cities Lab was designated to become a center of excellence for environmental remediation, and thus would de-emphasize the health and safety research that was being done there [USBM 1994].

While the recommendations for a reinvented Bureau were welcomed in some circles, the report drew strong criticism from the mining industry. The American Mining Congress sent a letter to the Bureau with comments on the reinvention report, opposing the large increase proposed in research on environmental technologies and reductions in the other research programs. The letter also stated that the mining industry was being ignored as a primary customer of the Bureau. It ended by recommending that the draft reinvention report be scrapped [Knebel 1994].

In 1990, the National Research Council (NRC) of the National Academies published a report on the competitiveness of the U.S. minerals and metals industry. This report also included sections on the role of Bureau research in the competiveness of the minerals industry. The following paragraph from the report describes the status of the Bureau within the federal system at that time:

> A basic problem in dealing with competition over agency responsibilities is the Bureau's location within the Department of the Interior, which has a traditional mandate to preserve and maintain public lands. Because the Bureau's mission involves mining,

which disturbs and exploits the land, its mission is somewhat at odds with other interests and concerns of the department. This situation apparently results in a lack of strong support for the Bureau in executive branch decisions and congressional hearings [National Research Council 1990].

To address this issue, the report recommended that the Bureau establish visiting committees to examine its research program on a more regular basis. As a response to this recommendation, the Bureau asked the NRC to assess the Bureau's research programs and resources and make recommendations to improve their quality. The NRC then established a Committee on Research Programs of the Bureau of Mines in early 1994. In its first year, the committee visited three Bureau labs and established a panel to conduct an in-depth review of the occupational health research program. The first report of the committee documented many of the problems that beset the Bureau and its research programs since the 1970s. The report noted that the Bureau had not been growing at the same rate as other federal research agencies in defense, aerospace, and energy:

> That this did not happen may have been due, in part, to the association of the bureau with an industry—mining—that had a worsening public image. Another reason, perhaps more valid, may be the historical placement of the bureau within the Department of the Interior (DOI). Long a leading department in peacetime, DOI in successive cold war administrations lost ground, not only to other existing cabinet functions but also to new departments and independent agencies. Taken together with the awkward relationship between the bureau's mission to assist in the exploitation of resources and DOI's charge to maintain, manage, and preserve public lands, a decline in the USBM was perhaps unavoidable [National Research Council 1994].

This report pointed out two major problems faced by the Bureau: (1) it served a primary customer, the minerals industry, which was generally viewed by the public as being of diminishing importance; (2) it was located in a department that viewed its mission increasingly as preservation of the environment on public lands, and was not favorable to a subordinate bureau that supported mining of those lands. Certainly, both of these issues caused the Department of the Interior to favor reductions in the Bureau budget.

The Bureau's reinvention project and the DOI budget proposed a smaller Bureau with program shifts toward environmental research. The fiscal year 1995 budget for the Bureau's health and safety research was to be reduced about 12 percent, while the budget for environmental research was to be increased by 25 percent. The overall budget reductions for fiscal year 1995 led the Bureau to plan the closure of its research laboratories at Rolla, MO, and Tuscaloosa, AL. The proposed consolidation of the research programs at the new centers of excellence caused research staff at the other centers to seek transfers, retirement, or employment elsewhere. The overall result was a loss of research staff and a decline in morale for many of the Bureau employees who remained.

A new threat to the Bureau arose in the 1994 Congressional election campaign, when the Republican "Contract with America" included as one of its proposals the termination of the Bureau along with several other agencies. The Republican Party took control of both houses of

Congress in the 1994 elections and implemented many of the provisions of the Contract with America. On July 18, 1995, the House of Representatives passed an appropriations bill, H.R. 1977, that provided funding only for the orderly closure of the Bureau. The Senate passed its own version of the appropriations bill, which included additional funding for the continuation of the Bureau. Finally, the House and Senate conference committee resolved their differences in September 1995 with a decision to close the Bureau and transfer some of its functions, staff, and facilities to other federal agencies. This decision became final with the passage of a continuing resolution for the fiscal year 1996 funding of the federal government.

At the time of its closure, the Bureau had facilities in 13 locations around the country in addition to its headquarters in Washington, DC. There were nine Research Centers, located in Albany, OR; Denver, CO; Minneapolis, MN; Pittsburgh, PA; Reno, NV; Rolla, MO; Salt Lake City, UT; Spokane, WA, and Tuscaloosa, AL. The Bureau also had three minerals information Field Operations Centers in Alaska (with offices in Anchorage and Juneau); Spokane, WA, and Denver, CO. In addition, there was a Minerals Availability Field Office in Denver, CO, and a Helium Operations facility in Amarillo, TX (Kirk, 1996).

Some provisions of the legislation that closed the Bureau were as follows:

- The mining research laboratory facilities and staff at Pittsburgh and Spokane and the health and safety research function were temporarily transferred to the Department of Energy. Later it was decided by the Office of Management and Budget (OMB) that these should be transferred to NIOSH.
- The materials research laboratory at Albany, OR was transferred to DOE.
- The mining environmental research being done at Pittsburgh and the associated staff were transferred to DOE and remained there.
- The minerals information function and part of the staff in Washington, DC, and Denver, CO, were transferred to the US Geological Survey at Reston, VA.
- The helium production function was transferred to the Bureau of Land Management for eventual closure in 1998.
- About 600 employees were involved in transfers to other agencies.
- The six Bureau research laboratories at Denver, Minneapolis, Reno, Salt Lake City, Rolla, MO, and Tuscaloosa, AL, were closed and the staff terminated. The Minerals Information offices in Denver and Spokane were also closed.
- The administrative functions and staff in Washington, DC, were terminated.
- More than 1,200 employees lost their jobs and the U.S. Bureau of Mines ceased to exist in 1996.

6. Consolidation and New Beginning: 1996-2010

6.1 Mining Research under NIOSH

The legislation that closed the U.S. Bureau of Mines (Public Law 104-99) also transferred the mining health and safety research function and associated staff and facilities at Pittsburgh and Spokane to the Department of Energy, with provision that a later decision should be made as to whether these should remain in DOE or be transferred to MSHA or NIOSH. The Fiscal Year

1996 Appropriations Bill stated in part that ". . . the health, safety and mining technology functions in Pittsburgh, PA and Spokane, WA will be continued under the Department of Energy. . ." and "The managers strongly encourage the Administration, and particularly the Office of Management and Budget, to work toward consolidating these health, safety, and mining technology functions in the same agency with either the Mine Safety and Health Administration or the National Institute for Occupational Safety and Health. . . ."

The Mining program was transferred to DOE early in 1996. Dr. Linda Rosenstock, the Director of NIOSH, had argued strongly for the transfer of the mining program to NIOSH, and a decision was made by the Office of Management and Budget (OMB) in February 1996 that NIOSH should acquire these functions. The Director of NIOSH and the Assistant Secretary for Fossil Energy of DOE entered into a Memorandum of Agreement on June 20, 1996, assigning day-to-day management of the program to NIOSH during fiscal year 1996. The formal transfer of staff and facilities took place in October 1996 under authority of Public Law 104-134.

The fiscal year 1997 Conference bill and report H.R. Omnibus Consolidated Appropriations Act, 1977 (Public Law 104-208, approved 9/30/96) stated that "while NIOSH has had responsibility for occupational health and safety research aimed at industry in general, the Committee understands that many mine safety and health needs are either unique to mining or require mining-specific emphasis. The Committee, therefore, expects NIOSH to preserve the integrity of mine safety and health research unit of the Bureau so that the collective experience and expertise of that group can be maintained within NIOSH. To further ensure the maintenance of this unit and its mission, the Committee recommends that NIOSH move forward with establishing a new Associate Director for Mining Safety and Health Research who reports directly to the NIOSH Director." NIOSH has maintained a separate and distinct identity to the mine health and safety program since that time.

NIOSH decided to keep its newly acquired (former Bureau) mining program both separate and visible within the agency by placing it in a newly created Office of Mine Safety and Health Research. This Office was to be headed by an Associate Director for Mining, reporting to the Director of NIOSH. The Associate Director has the lead responsibility for the review of the post-Bureau health and safety research program, the development of the appropriate organizational structure for this program within NIOSH, and the development of the long-term vision and direction for mining health and safety research. The Associate Director also oversees and provides management for mine safety and health research within NIOSH and is responsible for interaction with key mining safety and health stakeholders from industry, labor, and other government agencies.

The first Associate Director for Mining was Dr. R. Larry Grayson, who accepted this position in 1997. He was succeeded by Dr. Lewis V. Wade (2000-2004) and later by Dr. Jeffrey L. Kohler (2004-present). The names of the mining facilities were changed under NIOSH to the Pittsburgh Research Laboratory (PRL) and the Spokane Research Laboratory (SRL). A total of 413 staff positions transferred to NIOSH with the mining program, 336 at Pittsburgh and 77 at Spokane.

In 2001, NIOSH established the National Personal Protective Technology Laboratory (NPPTL), located in Bruceton, PA. The mission of NPPTL is to prevent and reduce occupational disease, injury, and death for workers who rely on personal protective technologies, such as respirators, gloves, and hard hats. Customers of NPPTL research include miners, firefighters, emergency responders, and health care, agriculture, and industrial workers. The PRL project and staff that worked on improving the design, reliability, and use of the self-contained self-rescuer were transferred to NPPTL.

6.2 NIOSH Mining-Related Research Accomplishments

6.2.1 Permissible Exposure Limit/Recommended Exposure Limit for Coal Mine Dust

Relying substantially on NIOSH epidemiologic research up to that time, in 1995 NIOSH produced and delivered to MSHA a criteria document recommending a reduced exposure limit for coal mine dust, along with other preventive interventions intended to reduce the risk of coal workers' pneumoconiosis among the nation's coal miners [NIOSH 1995]. NIOSH research showed that the current permissible exposure limit (PEL) of 2 mg/m^3 for respirable coal mine dust is not fully protective of coal miners' health. Although MSHA has yet to propose a reduction of the PEL to the NIOSH REL (recommended exposure limit) of 1.0 mg/m^3, significant reductions have been achieved in the percentage of coal mine dust samples exceeding the MSHA PEL since the early 1970s, despite large increases in coal production.

6.2.2 Coal Workers' Pneumoconiosis (CWP)

Findings from NIOSH's ongoing Coal Workers' X-ray Surveillance Program (CWXSP) showed that the prevalence of pneumoconiosis in underground miners with 25 or more years mining tenure fell from 35 percent in the years 1973-1978 to four percent in 1997-1999. However, more recent observations indicate an apparent upswing in prevalence (see Figure 38). Even the potentially disabling advanced form of the disease known as progressive massive fibrosis (PMF) continues to occur.

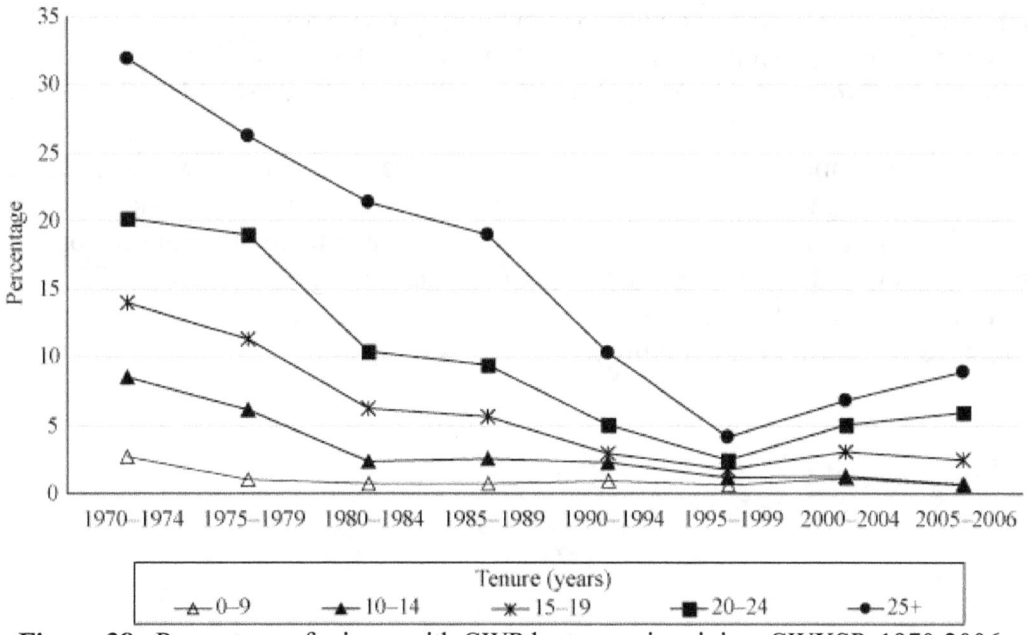

Figure 38. Percentage of miners with CWP by tenure in mining, CWXSP, 1970-2006.

Novel approaches undertaken by NIOSH staff have recently identified counties in the coalfields with high rates of rapid pneumoconiosis progression among active miners [Antao et al. 2005, 2006]. The reasons for this apparent rapid progression and continuing occurrence of progressive massive fibrosis are not entirely clear. However, they may result from a combination of inadequacies in the present dust limit and its method of enforcement. Certainly, these occurrences point to the need for continued monitoring activities to track disease and for continued research on enhanced exposure assessment and dust control.

6.2.3 Personal Dust Monitor (PDM)

In 1996, the Secretary of Labor asked NIOSH to develop a better dust monitoring instrument for coal mining. At that time, the coal mine personal sampler, a rugged and reasonably accurate instrument, had been in use for about 30 years. Nevertheless, because the internal filter must be weighed in a lab, there was always a delay of a week or more in getting measurement results back to the mine operator. Meanwhile, the miners themselves seldom had any knowledge of their dust exposure measurements. This delay meant that miners could be exposed to excessive dust levels without having the chance to immediately correct the problem and avoid overexposures.

Initial work to develop a better dust monitoring instrument focused on a machine-mounted area-sampling monitor, but NIOSH found this to be inadequate for compliance sampling of miners' dust exposure. Further, NIOSH findings showed no consistent relationship between personal sampling equipment worn by workers and area samples.

The work was then redirected in 1999 to develop a mass-based, continuous dust monitor that was small and light enough to be worn by the workers. Called a personal dust monitor, or PDM, it retained most of the features of the previous machine-mounted area monitor. One of the most important of these features was its ability to read out the gravimetric dust level during the work shift and to predict the full-shift dust level if nothing changed in the mining situation. This allowed workers and mine management to immediately correct dust controls to lower dust levels, thereby avoiding overexposure of the workers. At the end of the shift, the PDM data could be downloaded into a computer and stored for future reference. The concept was to empower miners and mine management to act on possible dust overexposures in real time. Conveniently, the dust monitor was built into a miner's cap lamp unit so that miners were putting on their dust monitors when they put on their cap lamps. The PDM underwent an extensive field evaluation to assess accuracy, reliability, and acceptance by mine workers, and results of the field evaluation were very positive (see Figure 39).

Figure 39. Miner wearing personal dust monitor (PDM).

In March 2003, MSHA published proposed changes to its dust sampling program. In these new regulations, MSHA allowed for the use of the continuous PDM. However, industry and labor indicated in public meetings that MSHA should delay rulemaking until the PDM could be further evaluated as a compliance-grade instrument. Both industry and labor representatives envision the PDM as having the potential to become the cornerstone for a new sampling process. Thus, a change in regulations has been proposed by MSHA and NIOSH to allow the certification of the PDM or any future similar dust monitor for use in determination of compliance with coal mine dust standards.

The PDM received an R&D 100 award as one of the nation's top 100 technical developments of the year in 2004. A PDM Coal Partnership has been established with members from industry, labor, and NIOSH to promote the PDM and to exchange information on the technology. In September 2008, the PDM received approval from MSHA for use in underground coal mines. Thermo-Fischer Scientific is now manufacturing the PDM and it is commercially available.

In addition to its use as a dust monitor, the PDM sampler has been modified to be used as a real-time diesel particulate monitor. The PDM unit is fitted with a particle size separator so that oversize material does not enter the unit and only diesel particulate matter is measured. Substantial testing of the modified PDM has been conducted in Australia, with promising results.

6.2.4 Improved Regulations for Dust Control Associated with Highwall Drilling at Surface Coal Mines

A series of NIOSH investigations documenting a very high risk of silicosis among rock drillers at surface coal mines was initially motivated by a fatal case of acute silicosis diagnosed in 1979. That miner, whose only silica exposure was 5 years as a highwall driller at a surface coal mine operation, died in his mid-30s as a direct result of his occupational exposure. Therefore, NIOSH researchers reanalyzed data from the earlier ALFORD survey of surface coal mines from the early 1970s, looking specifically at drill crew members and excluding those with prior underground experience. Pneumoconiosis prevalence among drill crew members was found to increase from 1.6 percent for those with less than ten years of drilling tenure to 15 percent for those with 20 or more years' experience. The NIOSH researchers concluded that "The belief that only coalminers who work underground have an appreciable risk of developing pneumoconiosis needs to be modified" [Banks et al. 1983].

In the mid-1980s, NIOSH carried out a new survey of the respiratory health status of surface coal miners [Amandus et al. 1989]. The survey found that pneumoconiosis was more prevalent among drillers (16 percent) than among others who had no drilling experience (≤ 3 percent). Prevalence among drillers increased with drilling tenure (up to 56 percent among those with more than 20 years drilling), and all cases of more advanced pneumoconiosis were found only in miners who had drilling experience. NIOSH also analyzed MSHA sampling data for surface coal mines across the U.S. for 1982-83 [Amandus and Piacitelli 1987], finding that excessive exposures to quartz were more frequent among drill crew members (> 80 percent) than among surface miners in other jobs.

As part of this same effort, NIOSH later analyzed MSHA data from 1982-86 [Piacitelli et al. 1990], finding that over 75 percent of the quartz samples for drillers exceeded the respirable silica PEL. In a listing of the ten jobs with the highest average respirable quartz exposures, the first three jobs were "highwall drill helper," "highwall drill operator," and "rockdriller," and the average concentration for these three together was double that of the next highest job on the list ("bulldozer operator").

NIOSH increased its efforts on this issue, developing engineering controls and other recommendations to reduce occupational silica exposures, e.g., improvements to dust collector on drills used by highwall drill operators, and cab-enclosure filtration systems. NIOSH had earlier produced an *Alert* warning employers and employees of the severe risk of improper operation of drilling equipment [NIOSH 1992].

With a collaborative NIOSH-MSHA-OSHA "Silica: It's not just dust!" prevention program in the 1990s, MSHA undertook enhanced efforts to reduce occupational exposures to silica and eliminate the incidence of silicosis, carrying out special industry-wide inspection "sweeps" of surface mines and establishing improved regulations for more effective enforcement of dust control on drilling rigs at surface mines. These new regulations required effective drill dust control regardless of dust exposure level [30 CFR § 72.620].

6.2.5 Diesel Particulate Regulations

Underground miners have routinely been exposed to very high concentrations of diesel particulate matter (DPM), much higher than any other occupation. In 1998, MSHA proposed two diesel rules—one for coal mining, the other for metal/nonmetal mining. The coal rule set a staged implementation limit on DPM emission ending at 2.5 g/hr for heavy-duty equipment, and this rule could be met by the use of MSHA-approved filters in the exhaust system. The metal rule set an interim concentration limit of 400 μg/m3 of total carbon and a final concentration limit of 160 μg/m3 of total carbon scheduled to go into effect in January 2006.

In 1999, a coal diesel partnership was formally established among union, industry, and NIOSH to work together to address technology to control diesel emissions in coal mines. The partners requested that NIOSH provide a review of available technology. In 2002, the metal/nonmetal industry, observing the coal partnership successes, requested that NIOSH establish a similar metal/nonmetal partnership to address diesel issues of concern in its own industry.

MSHA promulgated the new rule regulating underground metal and nonmetal miners' exposure to DPM in 2001. The rule specified that compliance would be determined by sampling a surrogate for DPM. The surrogate was to be the concentration of total carbon (TC) measured by using a size-selective impactor sampler. Total carbon is the sum of organic carbon (OC) and elemental carbon (EC). However, mining industry officials challenged the rule, claiming that sources of OC in the mine other than DPM, such as carbonaceous ore dust, tobacco smoke, and oil mist from drills, would falsely elevate TC above that attributable to DPM. The resulting settlement agreement among industry, labor, and MSHA specified that a metal/nonmetal mine study would be conducted in a variety of mines around the country, with NIOSH involved in the study.

Working together, NIOSH, MSHA, and the mining industry developed a protocol to investigate possible non-DPM sources of OC in underground metal/nonmetal mines and to evaluate the impact of those interferences on TC measurement. The mine study verified that drill oil mist and cigarette smoke could be significant sources of OC. The study also concluded that the non-DPM contributions to OC from oil mist and cigarette smoke could not be avoided, nor could an appropriate correction factor be applied to account for these OC sources.

As a result of this NIOSH research, MSHA, industry, and labor agreed to change the compliance surrogate for the DPM interim standard to EC, with EC being 77 percent of TC. The contested rule had incorporated the commercial version of the USBM diesel sampler to separate out carbon-bearing dusts from diesel soot, using total carbon (TC) as the DPM surrogate. NIOSH worked with MSHA and the manufacturer to improve the reliability and accuracy of the commercialized version of the USBM diesel sampler. It is now the MSHA-required sampler for sampling DPM in metal/nonmetal mines.

6.2.6 Epidemiologic Studies of the Vermiculite Mine at Libby, MT

In 1979, OSHA asked NIOSH to evaluate an unusual cluster of workers with bloody-pleural effusions at a fertilizer plant in Ohio. Airborne asbestos fibers were identified in the plant, which used expanded vermiculite as a carrier for fertilizer chemicals in a popular line of lawn products marketed nationwide. NIOSH investigators concluded that the asbestos was the likely cause of the pleural effusions and that the vermiculite used at this plant contained actinolite-tremolite asbestos and other closely related asbestiform amphibole fibers. The source of this vermiculite was a mine near Libby, Montana. Further, the potential for widespread exposures of both workers and consumers was readily apparent, given the wide usage of this vermiculite as loose-fill thermal insulation for attics, and in other materials used in construction, horticultural, and other applications. The Libby mine accounted for about three-fourths of worldwide vermiculite production for many decades.

After discovering the source of the asbestos-containing vermiculite and of NIOSH's case cluster finding in 1980, the Ohio fertilizer plant stopped using the vermiculite from Libby. NIOSH focused investigative efforts at the source of the hazard in Libby. To assess the risk for mine workers exposed to the asbestiform fibers, NIOSH undertook retrospective exposure assessment and morbidity and cohort mortality epidemiological studies. The main findings were as follows: fiber exposures for the vermiculite workers in Libby exceeded federal occupational exposure limits for asbestos; vermiculite workers at the Libby operations experienced substantial excess lung cancer mortality, which was dose-related to occupational fiber exposures at the vermiculite mine and associated processing facilities in Libby; pneumoconiosis and mortality from non-malignant respiratory diseases among vermiculite workers in Libby were also associated with occupational exposure to these fibers. When the studies were completed in the mid-1980s, NIOSH communicated these findings to MSHA, OSHA, and to local company management, union representatives, workers, and other interested members of the community. A series of publications followed in 1987 [Amandus et al. 1987a,b,c]. Finally, the mine at Libby was permanently closed in 1990.

In late 1999, a series of articles in a Seattle newspaper brought renewed national attention to concerns over community exposures in Libby and elsewhere. EPA and the Agency for Toxic Substances and Disease Registry (ATSDR) took leading roles at the federal level in assessing, communicating, and managing potential community health risks from asbestos exposures associated with the vermiculite, and NIOSH experts provided EPA and ATSDR with both formal and informal consultation. In addition, NIOSH has recently updated the mortality study of the Libby mining operations [Sullivan 2007] and provided formal testimony and comments for several Congressional hearings relating to Libby vermiculite and in support of MSHA's lowering of its PEL for asbestos in 2008 [30 CFR Parts 56, 57, and 71].

6.2.7 Reducing Noise on Mining Machines

MSHA data show that the continuous mining machine is first among all equipment in underground coal mining whose operators exceed 100 percent noise dosage. Continuous miners are large underground machines that cut coal at the working face of the mine. They gather up the cut coal and transport it via an onboard conveyor to the back of the machine. Here, it is loaded onto other mining equipment designed to carry the coal away from the working face. One of the main noise sources on a continuous mining machine is the onboard conveyor, which consists of a chain with flight bars that drags the coal along the base of the conveyor system. The metal chain and flight bars in contact with the metal base and the coal itself contribute greatly to the noise exposure of workers at the face. Continuous miner operators stand near the chain conveyor, at the back of the machine, and receive a great part of their noise exposure from the conveyor.

The urethane-coated flight bar chain was developed by NIOSH to reduce excessive noise exposure for operators of continuous mining machines. It has demonstrated an 8-hour Time

Weighted Average exposure reduction of 3 dB(A). The urethane coating also extended the chain life by preventing any chain link failures during the 6-month test period.

The dual-sprocket chain is another technology that was developed as a noise control for reducing noise overexposures of continuous mining machine operators. This control reduces noise by maintaining a constant level of tension and by decreasing chain slack that otherwise produces high-intensity noise-generating flight bar impacts at the conveyor transition points. Underground test results showed a 27 percent noise exposure reduction for the continuous mining machine operator.

Both the dual-sprocket chain and the urethane-coated flight bar chain were categorized on the MSHA Program Information Bulletin (PIB) P08-12 as "technologically and administratively achievable" [MSHA 2008]. Both these noise controls are now available commercially.

The Collapsible Drill Steel Enclosure (CDSE) was developed to reduce operator exposure to drilling noise from roof bolting machines. Drilling is the loudest noise source to which a roof bolting machine operator is typically exposed, contributing greatly to the operator's noise exposure. NIOSH testing determined that the drill steel and chuck radiate a significant amount of noise during drilling.

To address this problem, NIOSH developed the CDSE to encapsulate the drill steel during the drilling portion of the roof bolting machine duty cycle. Using a bellows to provide a barrier between the noise source and operator, the CDSE breaks the noise path between the drill steel and the operator (see Figure 40). Using a CDSE reduced the operator drilling noise levels by 6 to 7 dB(A) at the operator position of the machine. NIOSH is collaborating with J.H. Fletcher & Co. in manufacturing the CDSE as a standard component of the roof bolting machine.

Figure 40. Collapsible Drill Steel Enclosure at start of drilling.

6.2.8 Ground Control

Ground falls in U.S. underground mines caused more than 50,000 deaths or about half of all mining fatalities during the 20th century. Over the past 10 years, significant strides in ground control safety have been made. Important NIOSH ground control research products that have been successfully transferred to and implemented by the mining community during the past decade include:

- The Support Technology Optimization Program (STOP) and Analysis of Longwall Pillar Stability (ALPS) programs, which together have contributed to a virtual elimination of tailgate blockages.
- The underground stone ground control safety initiative. This initiative has increased awareness about rock fall hazards and helped to greatly reduce fatalities in stone mines.
- Guidelines for coal pillar recovery and widespread use of mobile roof supports. These have helped to make pillar recovery much safer.
- New types of standing roof supports. These have helped reduce the number of roof falls and the number of injuries to miners while installing roof support.
- A research and educational campaign aimed at increasing awareness about rock fall injuries and the use of surface controls in coal mines. This has helped to reduce rock fall injury rates.
- Mine design technologies such as the Coal Mine Roof Rating (CMRR) and Analysis of Horizontal Stress in Mines (AHSM), and guidelines for preventing massive pillar collapses. These have helped to provide more stable mining environments.

International collaboration is another measure of the caliber and impact of the NIOSH ground control program. During the past 8 years, six major funded research projects in Australia, Canada, the European Union, and the Republic of South Africa were based on NIOSH research. These focused on the STOP, ALPS, Analysis of Retreat Mining Pillar Stability (ARMPS), and CMRR software packages.

6.2.9 Coal Mine Roof Rating System

One reason that roof falls have been so difficult to eradicate is that the structural integrity of a coal mine's roof is greatly affected by natural geological weaknesses, including bedding planes, fractures, and small faults. The engineering properties of rock cannot be quantified solely by lab tests, because the strength of a small specimen is representative only of that sample, not taking into account the imperfections caused by local discontinuities.

NIOSH developed the Coal Mine Roof Rating (CMRR) system in the 1990s to fill the gap between geologic characterization and engineering design. This system combines many years of geologic studies in underground coal mines with worldwide experience with rock mass classification systems. New CMRR procedures for the borehole drill core, together with a software package, were developed in 2000 and made it possible to routinely collect CMRR data during geologic exploration. NIOSH has responded to requests for assistance in using the CMRR from many mining companies and geologic consultants, both domestically and abroad.

Worldwide experience has shown that the CMRR is a reliable, meaningful, and repeatable measure of mine roof quality.

NIOSH has developed a wide variety of ground control design tools based on the CMRR. These tools address a broad range of issues, including longwall pillar design, roof support selection, feasibility studies, extended-cut evaluation, and others. As a result, the CMRR is becoming the accepted standard for geotechnical characterization of mine roof and is often referred to in the technical literature. The CMRR has also become truly international, involved in mine designs and research projects funded by the Safety in Mines Research Advisory Committee (South Africa), the Canada Centre for Mineral and Energy Technology, and the Australian Coal Association Research Program.

6.2.10 Guidelines for Coal Pillar Recovery

When coal is first mined, large pillars of coal are left to support the rock between the mine and the surface. When these pillars are later recovered, the ground collapses. Nationally, coal pillar recovery accounts for just 10 percent of coal mined underground, but it is associated with more than 30 percent of mine roof fall fatalities.

NIOSH has conducted research to reduce the ground fall hazard during coal pillar recovery since the early 1990s. Significant research products include:

- The Analysis of Retreat Mining Pillar Stability (ARMPS) computer program (1994-1997)
- Advocacy of the use of mobile roof supports for temporary roof support (1994-present)
- Guidelines for coal pillar and panel design to prevent massive pillar collapses (1993-1997)
- Guidelines for panel and barrier pillars for pillar recovery under deep cover (2002)
- Guidelines for sizing the final stump to prevent unplanned roof collapse (2001)
- Guidelines for cut sequencing and roof bolting, and identification of other risk factors for pillar recovery, such as old works, multiple-seam mining, and wide roof spans (2002)

The research has also contributed to significant changes in the way pillar recovery is practiced throughout the United States. Today, more than one-half of all pillars are recovered using mobile roof supports. Fewer and fewer roof control plans allow the extraction of the final stump. Mine operators, MSHA, and state regulators are more aware of the risk factors and often specify extra support in pillar recovery sections.

6.2.11 Support Technology Optimization Program (STOP) Design Software

New mine roof support products with distinct performance characteristics are developed each year. The performance characteristics and limitations of these support systems are evaluated through full-scale testing in the NIOSH mine roof simulator at the Pittsburgh Research Laboratory. This information provides the basis for which these support systems are considered for specific applications by mining companies and approved for use by MSHA.

To facilitate the application of these technologies to improve mine safety, NIOSH developed the Support Technology Optimization Program (STOP). STOP is a Windows-based software program that provides mine operators with a simple and practical tool to make engineering decisions about the selection and placement strategy of these various mine roof support technologies. STOP can provide an engineering foundation to ensure that inadequate support designs, as well as ultraconservative support applications, are avoided. Safety is improved by matching support performance to mine conditions, reducing the likelihood of roof falls and blocked travel and escapeways.

Training workshops on the use of the STOP software have been held in eight states. More than 1,000 copies of the STOP software program have been distributed through these technology transfer seminars and from software download from the NIOSH webpage. The program has been used internationally in Mexico, the United Kingdom, Germany, Australia, and the Republic of South Africa.

6.3 Advancements in Mine Fires and Explosions Research

6.3.1 Coal Dust Explosibility Meter

Coal dust explosions are prevented by the addition of incombustible rock dust sufficient to render the coal dust inert. The Coal Dust Explosibility Meter (CDEM) is a hand-held instrument that provides instantaneous results of incombustible content instead of operators having to wait weeks for laboratory results from MSHA (see Figure 41). With real-time results, the potential for a disaster can be mitigated immediately. This meter can quickly determine the explosibility and the incombustible content of coal and rock dust mixtures in coal mines, thereby improving sample analysis and rock dusting practices.

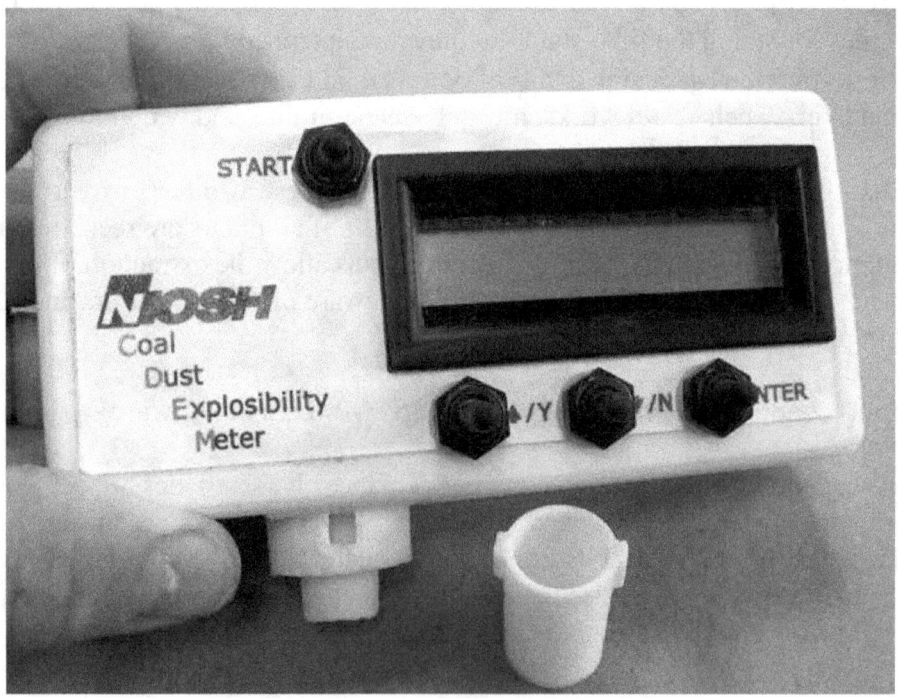

Figure 41. Coal Dust Explosibility Meter.

The CDEM measures the explosibility of a coal and rock dust mixture by an optical reflectance method. Since rock dust is white and coal dust is black, the intensity of the reflected light depends on the concentration of rock dust in the mixture. The CDEM, when calibrated with actual mine dust mixtures, can be used to determine the approximate percentage of rock dust in the coal and rock dust mixture.

The Pittsburgh Research Laboratory developed the theory and technology behind the CDEM, and a prototype was developed in collaboration with the Geneva College Center for Technology Development. NIOSH and partners received the prestigious R&D 100 Award for this instrument in 2006. In-mine testing showed that the CDEM results compared favorably with those from MSHA testing. With the CDEM initially developed under contract with a small prototyping firm, NIOSH is currently working with a manufacturer to commercialize this device for widespread distribution to the mining industry.

6.3.2 Improving Underground Coal Mine Sealing Strategies

Since 1986, 12 known explosions have occurred within the sealed areas of active U.S. underground coal mines. Most of these explosions destroyed the seals, structures built to isolate abandoned mining panels or groups of panels from the active workings. These explosions suggest that previously accepted seal construction and design methodologies are inadequate. One explosion resulted in the Sago disaster in which 12 miners died, while another caused the Darby disaster which claimed 5 lives. The new MSHA standard for sealing of abandoned areas has raised the explosion pressure design criteria for seals from the old standard of 20 psi to 50 psi or 120 psi, depending on whether or not the mining company chooses to monitor and maintain the sealed area atmosphere inert. This new standard is based in part on the recently published NIOSH report on "Explosion Pressure Design Criteria for New Seals in U.S. Coal Mines." This report established the scientific basis for these new seal standards based on the explosion pressures that could develop [Zipf et al. 2007].

6.3.3 Refuge Chambers

Refuge chambers are one option for safety protection in the event of explosions or fires. The Bureau described the advantages of refuge chambers in its 1912 report, but did not consider their installation to be as important as the prevention of mine explosions. Historically, in the U.S., the use of refuge chambers in coal mines has generated much debate. While refuge chambers can save lives, it is also argued that they may cause miners to seek refuge rather than attempt to escape their hazardous situation. A number of considerations are involved, including the capabilities of stations, the type and location of structures, design criteria, and maintenance and training issues.

To address these issues, section 13 of the Mine Improvement and New Emergency Response Act of 2006 (MINER Act) required NIOSH to conduct "research, including field tests, concerning the utility, practicality, survivability, and cost of various refuge alternatives in an underground coal mine environment, including commercially available portable refuge chambers." The resulting NIOSH report focused on specific information that could inform the regulatory process on refuge alternatives [NIOSH, 2007]. Further, gaps in knowledge and technology that should be addressed to help realize the full potential of refuge alternatives were also identified. The report concluded that refuge alternatives have the potential for saving lives of miners if part of a comprehensive escape and rescue plan, and if appropriate training is provided.

Two viable refuge alternatives are in-place shelters and portable chambers that are inflatable or rigid. In-place shelters provide a superior refuge environment and may possibly be connected to the surface by borehole. However, it would not be possible to keep them within 1000-2000 feet of the face, and extended distances would have to be allowed for them to be used.

NIOSH testing also found operational deficiencies in some commercially available portable chambers that could delay their deployment in mines. Testing and certification of both refuge chambers and in-place shelters would be needed. NIOSH concluded that while additional development is needed, the benefits of these refuge alternatives merit their deployment in underground coal mines.

The 2006 MINER Act also directed MSHA to consider rulemaking on refuge chambers. However, Congress subsequently mandated action in language included in the 2008 appropriations bill. MSHA proposed a regulation for three types of acceptable refuge alternatives: pre-fabricated self-contained units, a secure space that is constructed in place, or materials pre-positioned for miners to use to construct a secure space. MSHA issued the final rule for refuge alternatives for underground coal mines on December 31, 2008.

6.4 Advancements in Injury Prevention and Research

6.4.1 Hazardous Area Signaling and Ranging Device (HASARD)—Proximity Warning System

Based on the use of low-frequency magnetic fields to provide a danger marker around machinery, HASARD is a proximity warning system created for the purpose of warning continuous mining machine operators when they are in an unsafe location around continuous mining machines. Between 1984 and 2008, there were 30 crushing or pinning fatalities around continuous mining machines in underground coal mining. A collection of different proximity system concepts was investigated, with low-frequency magnetic fields showing the most promise.

As part of this research, many different prototypes were designed, built, and tested, both in the laboratory and at select local mines. The final design consisted of a magnetic field generator placed on the mining machine. A miner-worn receiver measures the magnetic field to determine when he is close to a dangerous area. An alarm is provided to the miner acknowledging the danger. If the miner gets too close, the continuous mining machine is shut down.

Two patents were acquired representing the technology, and three companies have since licensed the patents. Four products are now available on the market based on these licenses. One international CRADA continues to be supported to transfer the technology.

6.4.2 Improved Training Materials and Methods

Mine safety and health professionals have long recognized training as a critical part of an effective safety and health program. Since 1977, federal regulations have required mine operators to provide safety and health training to all new miners, as well as a minimum of 8 hours of refresher training each year. With help from universities, mining companies, MSHA, and other providers of miner training, NIOSH researchers developed more than 80 training modules and products on a wide variety of safety and health topics. Most of these are intended to improve miners' ability to (1) recognize common workplace hazards, or (2) handle non-routine events such as fires and other types of mine emergencies.

The main emphasis of the NIOSH Mining Program's training research activity is not on producing training materials per se, but on finding better training processes and methods. Most of these training modules were developed during research studies to determine the feasibility and effectiveness of using innovative new methods to present occupational safety and health information to miners. These include computer simulations, interactive problem-solving stories, degraded stereoscopic (3-D) images of hazardous conditions, and videotaped interviews with miners.

According to MSHA records, NIOSH's mine training materials have been used extensively by the mining industry. During the past 20 years, trainers have obtained more than a half million copies of training materials through MSHA's National Mine Health and Safety Academy. Numerous mining companies, mine trainers, and union officials have requested help from NIOSH mine training researchers. Several companies have provided financial support through Cooperative Research and Development Agreements. MSHA and state mining officials often request NIOSH help and advice on various matters related to miners' safety and health training.

6.5 Current NIOSH Mining Program

Today, NIOSH performs research to improve occupational safety and health for all workers, including miners. NIOSH assumed the sole federal responsibility for research on mine safety after the closure of the U.S. Bureau of Mines in 1996, when NIOSH inherited the former Bureau mining research laboratories in Pittsburgh, PA, and Spokane, WA. In 2009, the NIOSH Office of Mine Safety and Health Research had about 225 employees at these two locations, and a budget of $40.6 million.

As of 2009, the NIOSH mining program had seven major strategic goals:

1. Reduce respiratory diseases in miners associated with coal worker pneumoconiosis, silicosis, and diesel emissions;
2. Reduce noise-induced hearing loss (NIHL);
3. Reduce repetitive/cumulative musculoskeletal injuries;
4. Reduce traumatic injuries;
5. Reduce the risk of mine disasters (fires, explosions, and inundations) and enhance the safety and effectiveness of, emergency responders;

6. Reduce ground failure fatalities and injuries;
7. Determine the impact of changing mining conditions, new and emerging technologies, training, and the changing patterns of work.

6.5.1 Partnerships

NIOSH has created numerous partnerships to help meet high priority research needs in mining. These partnerships address the personal dust monitor, diesel exposures in coal and metal/nonmetal mines, hearing loss prevention, coal mine seals, and emergency communications. Besides MSHA, partners include the United Steelworkers, United Mine Workers, Operating Engineers, National Mining Association, Bituminous Coal Operators Association, West Virginia University, and the National Stone, Sand & Gravel Association.

6.5.2 NAS Evaluation of NIOSH Mining Program

Recently, NIOSH contracted with the National Academies of Science and Engineering (NAS) to conduct an evaluation of its research programs, including the Mining Program. Specifically, the NAS was tasked to evaluate the relevance and impact of the Program, and to assign a numerical score to represent its assessment. Additionally, the NAS was tasked to examine future issues and provide recommendations on areas for consideration of future research. The Mining Program prepared an "evidence package" to document its activities and impact over the past ten years, submitting it to the NAS in December 2005. Leadership from the Mining Program first met with the NAS review committee on January 11, 2006.

The NAS completed its review and presented its findings to NIOSH on April 27, 2007. The findings and recommendations of the committee are documented in its report entitled, "Mine Safety and Health Research at NIOSH" [National Research Council 2007]. The committee assessed the relevance and impact of the Mining Program, concluding the following:

> ...*research of the Mining Program is in high-priority areas and adequately connected to improvements in the workplace. A rating of 4 on a five-point scale (where 5 is the highest) is appropriate. Contributions of the program to improvements in workplace health and safety during the period evaluated (1997 to 2005) are considered major in some areas (respirable disease prevention, traumatic injury prevention), moderate in some areas (hearing loss prevention, ground failure prevention), and likely in a number of areas (disaster prevention, musculoskeletal injury prevention). Mining Program outputs are evaluated, accepted, and incorporated into stakeholder operations, and training outputs find wide use in the industry. The Mining Program is moderately engaged in technology transfer activities. A score of 4 for impact is appropriate.*

The report recognizes the many accomplishments of the NIOSH Mining Program over the past eight years. These scores of 4 (on a scale of 5) for relevance and 4 (on a scale of 5) for impact recognize that taxpayer resources are being focused in high-priority areas, and that over the past ten years NIOSH mining research has had significant impact on improving the health and safety of miners. The report also provides important insights into opportunities to further improve the relevance and impact of the Mining Program. NIOSH staff studied the NAS draft report and completed a response, including an action plan in December 2007, although implementation of many of the recommendations had already begun.

6.6 Recent External Factors Affecting Mining Research

Several recent coal mining disasters have had a major impact on mining in the United States. These incidents began in 2006 with an explosion on January 2 at the Sago Mine in West Virginia, killing 12 miners. A fire at the Aracoma Alma No. 1 Mine in West Virginia on January 19 caused 2 fatalities. On May 20, 2006, there was an explosion at the Darby No. 1 Mine in Kentucky, resulting in 5 more deaths. These mine disasters led Congress to pass the MINER Act on May 24, 2006. Some of the major provisions of the MINER Act include:

- A required wireless two-way communications and electronic tracking system within three years, permitting those on the surface to locate persons trapped underground;
- A requirement for each mine to have available two experienced rescue teams capable of a one-hour response time;
- Required new safety standards relating to the sealing of abandoned areas in underground coal mines, increasing the requirements for strength of the seals;
- Establishment of a competitive grant program for new mine safety technology to be administered by NIOSH;
- Establishment of a Technical Study Panel for Belt Air to investigate the use of air from conveyer belt mine entries to ventilate the mine face;
- Establishment of a permanent Office of Mine Safety and Health Research in NIOSH, helping to ensure a viable and long-term focus on mining safety and health.

The MINER Act also required MSHA to act on new regulations and established deadlines by which some of the new regulations were to be finalized.

On August 6, 2007, six miners were killed by a coal outburst at the Crandall Canyon Mine in Utah, when pillars failed and violently ejected coal over a large area. Ten days later, two mine employees and an MSHA mine inspector were killed in a coal outburst during rescue efforts. The Crandall Canyon mine disaster increased pressure in Congress for action to improve safety and health for American miners.

In January of 2008, the House of Representatives passed a bill called the Supplemental Mine Improvement and New Emergency Response Act of 2007, or the "S-MINER Act." The S-MINER Act would have very significant impact on the American mining industry if it were to become law, but the bill did not reach the floor of the Senate in 2008. The S-MINER Bill passed by the House of Representatives would have especially strengthened mining health regulations. It would have permitted the new continuous personal dust monitor (CPDM) to be used to determine compliance with the respirable coal mine dust standard. It would have cut the exposure limits for respirable coal mine dust and silica in half. Finally, it would have required MSHA to accept the scientific recommendations of NIOSH to update health hazard exposure limits.

7. Conclusion: Lessons Learned

The need to transfer research results to potential users has been clear since the establishment of the U.S. Bureau of Mines in 1910. Dr. Joseph A. Holmes placed initial emphasis on safety training for miners by the new Bureau. He used the mine rescue railroad cars to provide safety training to miners at the mine sites. The Bureau participated in exhibitions of mine safety technology, going back to the demonstration attended by President Taft in Pittsburgh in October 1911. Today, the Bureau continues to exhibit mine safety technology, frequently by way of trade shows, conferences, and professional society meetings (see Figure 42). The concept was called technology transfer by the Bureau in the 1970s, and is now called research to practice (r2p) by NIOSH, but the critical need to transfer new technology into use at the workplace has always been clear.

Figure 42. NIOSH Exhibit at the Society for Mining, Metallurgy, and Exploration Annual Meeting, Denver, 2009.

Mining research has been most effective when it has been done in cooperation with the ultimate users of the research. Successful transfer of research results requires close cooperation of researchers with staff at operating mines and field testing of new technology in those mines. Some mining companies have shown great initiative through the years in developing and testing new mining technologies to improve safety and health. A few companies have engaged in their own research activities, while others have been willing to work with the Bureau and NIOSH in allowing the use of their mines for developing and evaluating new technologies. This has often been done at a cost to the companies either directly, or through some loss in production. The work of the federal government in mining safety and health research has been very dependent on the cooperation of these mining companies and the people operating the mines, both in labor and management.

The mining industry has seen a mix in terms of early adopters of technology and very late adopters. This report has considered only the mining industry, so it is not possible to say here whether mining has been any better or worse than other industries in embracing new methods to improve worker safety and health. Historically, mining has been relatively early to work on improvements in safety technology because the need has been so apparent. Few other industries have had to face the complex and specialized problems of mining, including the dangers of roof falls, difficult ventilation challenges, gas and dust explosions, and limited worker escape routes. A century ago, mining was clearly the most dangerous industry in the United States. This fact led Congress to establish the U.S. Bureau of Mines in 1910 to address these problems.

Significant improvements in safety and health outcomes for miners often have resulted from changes in laws or regulations that forced adoption of safer technology. Improved technology also has been adopted voluntarily as the safety culture has evolved, and levels of injury and illness that were considered acceptable in the past are no longer tolerated. Research products have not necessarily been adopted quickly, particularly if they were perceived to add to the economic costs of mining. However, new and safer mining technology developments have eventually enabled the setting of better safety and health standards and adoption of a stronger safety culture in mining.

Reviewing the history of mining over the past century shows that each generation of miners has expected and demanded safer mining conditions than those that had been considered acceptable in past decades. Much should continue to be done to encourage miners and mine management toward a safety culture that looks for and expects continuous improvement in mining safety and health. Such a safety culture in mining will lead to faster adoption of improvements in mining technology.

There have been vast improvements in safety and health in the mining industry in the past 100 years. To demonstrate, Table 2 shows the numbers of fatalities in coal mines and Table 3 lists the numbers for metal and nonmetal mines in the years since 1900 in the U.S. In 2008, U.S. coal miners produced a new high of 1.17 billion tons of coal. There were 29 fatalities among the miners who produced that coal in 2008, a three-year low in fatal coal mining accidents [Coal Age 2009]. Changes in mining technology have led to great improvements in safety and health, while allowing for increased productivity. The research done by the Bureau and NIOSH has done much to enable the improvements that have been made in mining technology.

Unfortunately, there are still significant safety challenges remaining, as illustrated by the recent mine disasters in 2006 and 2007. From a health standpoint, the increased incidence of CWP found in coal miners in some areas of the country shows that health problems have not yet been solved. To meet these challenges, the NIOSH mining program will continue in the tradition of its predecessors to work with partners and stakeholders to conduct research, transfer it into practice, and track the impact of its efforts. We will continue to strive to make mines as safe and as healthy as any other workplace.

8. References

Amandus HE, Piacitelli G [1987]. Dust exposures at U.S. surface coal mines in 1982-1983. Arch Environ Health 42(6):374-381. Amandus HE, Petersen MR, Richards TB [1989]. Health status of anthracite surface coal miners. Arch Environ Health *44*(2):75–81.

Amandus HE, Wheeler R [1987]. The morbidity and mortality of vermiculite miners and millers exposed to tremolite-actinolite: Part II. Mortality. Am J Ind Med *11*(1):15–26 (A3–88).

Amandus HE, Althouse R, Morgan WK, Sargent EN, Jones R [1987a]. The morbidity and mortality of vermiculite miners and millers exposed to tremoliteactinolite: Part III. Radiographic findings. Am J Ind Med *11*(1):27–37.

Amandus HE, Wheeler R, Jankovic J, Tucker J [1987b]. The morbidity and mortality of vermiculite miners and millers exposed to tremolite-actinolite: Part I. Exposure estimates. Am J Ind Med *11*(1):1–14.

Antao VC, Petsonk EL, Sokolow LZ, Wolfe AL, Pinheiro GA, Hale JM, Attfield MD [2005]. Rapidly progressive coal workers' pneumoconiosis in the U.S.: Geographic clustering and other factors. Occup Environ Med *62*:670–674.

Antao VC, Petsonk EL, Attfield MD (Centers for Disease Control and Prevention) [2006]. Advanced cases of coal workers' pneumoconiosis, Lee and Wise counties, Virginia. MMWR *55*:909–913.

Banks DE, Bauer MA, Castellan RM, Lapp NL [1983]. Silicosis in surface coalmine drillers. Thorax. *38*(4):275–278.

Coal Age [2009]. Coal Age Magazine, January:2.

Coward HF, Jones GW [1952]. Limits of flammability of gases and vapors. Washington, DC: US Bureau of Mines, Bulletin 503.

CRS [1976]. The U.S. Bureau of Mines. Prepared by Congressional Research Service, Library of Congress, for the U.S. Senate, Committee on Interior and Insular Affairs.

DOI [1934]. Annual Report of the Secretary of the Interior, Fiscal Year Ended June 30, 1934. Washington, DC: Department of the Interior.

DOI [1935]. Annual Report of the Secretary of the Interior, Fiscal Year Ended June 30, 1935. Washington, DC: Department of the Interior.

DOI [1946]. Annual Report of the Secretary of the Interior, Fiscal Year Ended June 30, 1946. Washington, DC: Department of the Interior.

DOI [1948]. Annual Report of the Secretary of the Interior, Fiscal Year Ended June 30, 1948. Washington, DC: Department of the Interior.

DOI [1970].Towards improved health and safety for America's coal miners. Annual Report of the Secretary of the Interior. Washington, DC: Department of the Interior.

DOI [1973]. Administration of the Federal Coal Mine Health and Safety Act, Annual Report of the Secretary of the Interior. Washington, DC: Department of the Interior.

Doyle HN [2009]. The Federal Industrial Hygiene Agency: A History of the Division of Occupational Health, U.S. Public Health Service. Prepared for the History of Industrial Hygiene Committee, American Conference of Governmental Industrial Hygienists. NIOSH Intranet site, http://mtn.niosh.cdc.gov/eid/rmca/homepage/rmca/doyle.html, accessed June 3, 2009.

Fay AH [1916]. Coal mine fatalities in the United States, 1870–1914. U.S. Bureau of Mines, Bulletin 115.

Higgins E, Lanza AJ, Laney FB, Rice GS [1917]. Siliceous dust in relation to pulmonary disease among miners in the Joplin District, Missouri. US Bureau of Mines, Bulletin 132.

Humphrey, HB [1960]. Historical summary of coal-mine explosions in the United States, 1810–1958. US Bureau of Mines Bulletin 586.

Kaas LM [2006]. A history of the U.S. Bureau of Mines, with some highlights of its involvement in anthracite mining. Thirteenth Annual Journal of The Mining History Association, pp. 73–92.

Kirk WS [1996]. The history of the Bureau of Mines, Reprinted from USBM Minerals Yearbook, 1994.

Knebel JA [1994] Letter from John A Knebel, President, American Mining Congress, to Dr. Herman Enzer, Acting Director of the U.S. Bureau of Mines. January 19, 1994.

MSHA [2008]. Program Information Bulletin (PIB) P08–12. U.S. Mine Safety and Health Administration.

National Research Council [1990]. Competitiveness of the U.S. minerals and metals Industry, p. 109.

National Research Council [1994]. Research Programs of the U.S. Bureau of Mines, First assessment, p. 6.

National Research Council [2007]. Mining Safety and Health Research at NIOSH: reviews of research programs of the National Institute for Occupational Safety and Health. Committee to Review the NIOSH Mining Safety and Health Research Program, Committee on Earth Resources, National Research Council. The National Academies Press, Washington, DC.

National Research Council and Institute of Medicine [2007]. Mine Safety and Health Research at NIOSH, National Academies Press.

NIOSH [1992]. NIOSH Alert: Preventing silicosis and deaths in rock drillers. Cincinnati, OH: U.S. Department of Health and Human Services, Centers for Disease Control, National Institute for Occupational Safety and Health, DHHS (NIOSH) Publication No. 92–107.

NIOSH [1995]. Criteria for a Recommended Standard: occupation exposure to respirable coal mine dust. Cincinnati, OH: U.S. Department of Health and Human Services, Centers for Disease Control and Prevention, National Institute for Occupational Safety and Health, DHHS (NIOSH) Publication No 95–106.

NIOSH [2007]. Research report on refuge alternatives for underground coal mines. Report prepared in response to Section 13 of the MINER Act of 2006.

Piacitelli GM, Amandus HE, Dieffenbach A [1990]. Respirable dust exposures in U.S. surface coal mines (1982–1986). Arch Environ Health *45*(4):202–209.

Powell, FW [1922]. The Bureau of Mines—its history, activities and organization. Institute for Government Research.

Rice GS [1911]. The explosibility of coal dust. US Bureau of Mines, Bulletin 20.

Rice GS, Jones LM, Clement JK, Egy WL [1913]. First series of coal dust explosion tests in the experimental mine. US Bureau of Mines, Bulletin 56.

Rice GS, Jones LM, Egy WL, Greenwald HP [1922]. Coal-dust explosion tests in the experimental mine 1913 to 1918, inclusive. US Bureau of Mines, Bulletin 167.

Sullivan PA [1990]. Vermiculite, respiratory disease and asbestos exposure in Libby, Montana: update of a cohort mortality study. Environ Health Perspect *115*(4):579–85.

Tuchman RJ, Brinkley RF [1990]. A history of the Bureau of Mines Pittsburgh Research Center. US Department of the Interior, Bureau of Mines.

Turner S [1934]. The United States Bureau of Mines today. Mining Congress J, January:35–36.

USBM [1911]. First Annual Report of the Director of the Bureau of Mines to the Secretary of the Interior for the Fiscal year Ended June 30, 1911. Washington DC: Government Printing Office.

USBM [1912]. Second Annual Report of the Director of the Bureau of Mines to the Secretary of the Interior for the Fiscal year Ended June 30, 1912. Washington DC: Government Printing Office.

USBM [1913]. Third Annual Report of the Director of the Bureau of Mines to the Secretary of the Interior for the Fiscal year Ended June 30, 1913. Washington DC: Government Printing Office.

USBM [1914]. Fourth Annual Report of the Director of the Bureau of Mines to the Secretary of the Interior for the Fiscal year Ended June 30, 1914. Washington DC: Government Printing Office.

USBM [1915]. Fifth Annual Report of the Director of the Bureau of Mines to the Secretary of the Interior for the Fiscal year Ended June 30, 1915. Washington DC: Government Printing Office.

USBM [1925]. Fifteenth Annual Report of the Director of the Bureau of Mines to the Secretary of the Interior for the Fiscal year Ended June 30, 1925. Washington DC: Government Printing Office.

USBM [1932]. Annual Report of the Director of the Bureau of Mines to the Secretary of Commerce for the Fiscal year Ended June 30, 1932. Washington DC: Government Printing Office.

USBM [1933]. Annual Report of the Director of the Bureau of Mines to the Secretary of Commerce for the Fiscal year Ended June 30, 1933. Washington DC: Government Printing Office.

USBM [1960]. Annual Progress Report for Fiscal Year 1960, U.S. Department of the Interior, Bureau of Mines.

USBM [1965]. Research Facilities Program to the Year 2000. US Department of the Interior, Bureau of Mines.

USBM [1969]. Annual Progress Report for Fiscal Year 1969, U.S. Department of the Interior, Bureau of Mines.

USBM [1973]. Annual Progress Report for Fiscal Year 1973. U.S. Department of the Interior, Bureau of Mines.

USBM [1993]. Mine health and safety research, achievements of the U.S. Bureau of Mines. U.S. Department of the Interior, U.S. Bureau of Mines.

USBM [1994]. Reinventing the USBM. (Draft document that was not published.)

Zipf RK, Sapko MJ, Brune JF [2007]. Explosion pressure design criteria for new seals in U.S. Coal Mines. Pittsburgh, PA: U.S. Department of Health and Human Services, Centers for Disease Control and Prevention, DHHS (NIOSH) Publication No. 2007–144, Information Circular 9500.

Appendix A: Joseph A. Holmes Biography

Joseph Austin Holmes, the first Director of the U.S. Bureau of Mines (USBM), was born at Laurens, SC, on November 23, 1859. He graduated from Cornell University with the degree of Bachelor of Agriculture in 1881. He was professor of geology and natural history at the University of North Carolina during 1882-91. Holmes resigned the professorship in 1891 to become the State Geologist with the North Carolina Geological Survey.

Geological themes were the topics of most of Holmes's nearly eighty publications. In 1904, the President appointed Dr. Holmes chief of the laboratories for testing fuel and structural materials at the United States Geological Survey (USGS). Subsequently, he became chief of the technological branch of the USGS. He was appointed by President Taft as the Director of the USBM upon its formation in 1910. Under his management of the Bureau, new methods were perfected for preventing mine accidents.

Dr. Holmes established the experimental mine for testing explosions at Bruceton, PA, and in October 1911, organized the first national mine safety demonstration. He was responsible for the establishment of federal and state rescue stations in the coal and metal mining regions and the equipping of railroad cars as mobile safety and rescue stations. He was a fellow and charter member of the Geological Society of America and a fellow of the American Association for the Advancement of Science. He was a member of the Academies of Science in Washington, DC; the American Forestry Association; the American Institute of Mining Engineers; the American Society for Testing Materials; and the American Society of Mechanical Engineers.

When Dr. Holmes retired in 1915, his health had been declining. He died on July 12, 1915, at the age of fifty-five from pulmonary tuberculosis in Denver, Colorado. He is buried in Rock Creek Cemetery, Washington, DC [Kirk 1996].

The Joseph A. Holmes Safety Association (JAHSA) was founded in 1916, the year after the death of Dr. Holmes. Twenty-four leading national organizations representing the mining, metallurgical, and allied industries aimed to promote health and safety in the mining industry by creating a safety awards program. Various individual awards such as the hero award and merit award are issued by the Association, as well as company awards for time worked without accidents or fatalities. The association is a nonprofit organization consisting of representatives from federal and state governments, mining organizations, and labor.

Appendix B: Directors of USBM and NIOSH

The USBM had 19 permanent directors.

USBM Permanent Directors

Joseph A. Holmes	1910-15
Van H. Manning	1915-20
Frederick G. Cottrell	1920
H. Foster Bain	1921-25
Scott Turner	1925-34
John W. Finch	1934-40
Royd R. Sayers	1940-47
James Boyd	1947-51
John J. Forbes	1951-55
Marling J. Ankeny	1956-64
Walter J. Hibbard, Jr	1964-68
John F. O'Leary	1968-70
Elburt F. Osborn	1970-73
Thomas V. Falkie	1974-77
Roger A. Markle	1978-79
Lindsay D. Norman, Jr	1980-81
Robert C. Horton	1981-87
T S Ary	1988-93
Rhea L. Graham	1994-96

NIOSH Permanent Directors

Marcus M. Key	1971-75
John F. Finklea	1975-78
Anthony Robbins	1979-81
J. Donald Millar	1981-93
Linda Rosenstock	1994-00
John Howard	2002-08
John Howard	2009-p

Table 1. U.S. Bureau of Mines appropriations and expenditures, fiscal years ending June 30, 1911-35

Fiscal year	Appropriated to U.S. Bureau of Mines	Total expenditures
1911	$502,200.00	$513,581.73
1912	475,500.00	514,900.23
1913	583,100.00	626,862.80
1914	664,000.00	716,629.50
1915	730,500.00	781,740.49
1916	757,300.00	796,952.24
1917	981,060.00	984,871.90
1918	1,467,070.00	4,185,226.88
1919	3,245,285.00	9,442,591.08
1920	1,216,897.00	1,200,104.82
1921	1,362,642.00	2,077,994.83
1922	1,474,300.00	1,664,179.55
1923	1,580,900.00	1,737,855.22
1924	1,784,959.00	2,145,403.57
1925	2,028,268.00	2,214,490.66
1926	1,875,010.00	2,437,839.37
1927	1,914,400.00	2,288,972.10
1928	3,025,150.00	2,730,180.83
1929	2,725,118.00	3,600,393.33
1930	2,274,670.00	2,548,671.45
1931	2,745,060.00	2,939,060.73
1932	2,278,765.00	2,426,022.75
1933	1,860,325.00	1,910,612.04
1934	1,574,300.00	1,481,497.72
1935	1,293,959.07	1,511,047.76

Note that these amounts have not been adjusted for inflation and would be much higher in current dollars. The budget figures are for the entire U.S. Bureau of Mines and not just for safety and health research.

Table 2. U.S. Coal Fatalities, 1900 through 2008.

Year	Miners	Fatalities	Year	Miners	Fatalities	Year	Miners	Fatalities	Year	Miners	Fatalities
1900	448,581	1,489	1930	644,006	2,063	1960	189,679	325	1990	168,625	66
1901	485,544	1,574	1931	589,705	1,463	1961	167,568	294	1991	158,677	61
1902	518,197	1,724	1932	527,623	1,207	1962	161,286	289	1992	153,128	55
1903	566,260	1,926	1933	523,182	1,064	1963	157,126	284	1993	141,183	47
1904	593,693	1,995	1934	566,426	1,226	1964	150,761	242	1994	143,645	45
1905	626,045	2,232	1935	565,202	1,242	1965	148,734	259	1995	132,111	47
1906	640,780	2,138	1936	584,582	1,342	1966	145,244	233	1996	126,451	39
1907	680,492	3,242	1937	589,856	1,413	1967	139,312	222	1997	126,429	30
1908	690,438	2,445	1938	541,528	1,105	1968	134,467	311	1998	122,083	29
1909	666,552	2,642	1939	539,375	1,078	1969	133,302	203	1999	114,489	35
1910	725,030	2,821	1940	533,267	1,388	1970	144,480	260	2000	108,098	38
1911	728,348	2,656	1941	546,692	1,266	1971	142,108	181	2001	114,458	42
1912	722,662	2,419	1942	530,861	1,471	1972	162,207	156	2002	110,966	28
1913	747,644	2,785	1943	486,516	1,451	1973	151,892	132	2003	104,824	30
1914	763,185	2,454	1944	453,937	1,298	1974	182,274	133	2004	108,734	28
1915	734,008	2,269	1945	437,921	1,068	1975	224,412	155	2005	116,436	23
1916	720,971	2,226	1946	463,079	968	1976	221,255	141	2006	122,975	47
1917	757,317	2,696	1947	490,356	1,158	1977	237,506	139	2007	122,936	34
1918	762,426	2,580	1948	507,333	999	1978	255,588	106	2008	133,827	30
1919	776,569	2,323	1949	485,306	585	1979	260,429	144			
1920	784,621	2,272	1950	483,239	643	1980	253,007	133			
1921	823,253	1,995	1951	441,905	785	1981	249,738	153			
1922	844,807	1,984	1952	401,329	548	1982	241,454	122			
1923	862,536	2,462	1953	351,126	461	1983	200,199	70			
1924	779,613	2,402	1954	283,705	396	1984	208,160	125			
1925	748,805	2,518	1955	260,089	420	1985	197,049	68			
1926	759,033	2,234	1956	260,285	448	1986	185,167	89			
1927	759,177	2,231	1957	254,725	478	1987	172,780	63			
1928	682,831	2,176	1958	224,890	358	1988	166,278	53			
1929	654,494	2,187	1959	203,597	293	1989	164,929	68			

Total number of coal mining fatalities from 1900 through 2008: **104,656.**

Note: Data on office workers is included starting in 1973.

Data from MSHA website (10/7/09) at http://www.msha.gov/stats/charts/chartshome.htm

Table 3. Metal/Nonmetal Fatalities, 1900 through 2008.

Year	Miners	Fatalities	Year	Miners	Fatalities	Year	Miners	Fatalities	Year	Miners	Fatalities
1900	N/A	N/A	1930	N/A	376	1960	289,001	185	1990	235,690	56
1901	N/A	N/A	1931	159,007	225	1961	279,178	127	1991	230,107	53
1902	N/A	N/A	1932	116,079	139	1962	269,927	216	1992	224,567	43
1903	N/A	N/A	1933	125,347	159	1963	259,926	173	1993	219,320	51
1904	N/A	N/A	1934	138,689	181	1964	260,939	179	1994	225,498	40
1905	N/A	N/A	1935	177,160	222	1965	263,072	180	1995	229,536	53
1906	N/A	N/A	1936	193,957	308	1966	261,993	195	1996	229,045	47
1907	N/A	N/A	1937	217,020	310	1967	255,999	181	1997	235,915	61
1908	N/A	N/A	1938	192,567	247	1968	246,039	182	1998	235,561	51
1909	N/A	N/A	1939	203,843	232	1969	246,677	179	1999	238,852	55
1910	N/A	N/A	1940	213,619	307	1970	242,788	165	2000	240,450	47
1911	N/A	883	1941	226,220	322	1971	237,059	164	2001	232,770	30
1912	N/A	874	1942	210,409	361	1972	185,115	234	2002	218,148	42
1913	N/A	866	1943	183,726	314	1973	246,665	175	2003	215,325	26
1914	N/A	739	1944	158,282	237	1974	271,606	158	2004	220,274	27
1915	N/A	701	1945	145,637	174	1975	277,978	123	2005	228,401	35
1916	N/A	870	1946	162,408	181	1976	278,605	113	2006	240,522	26
1917	N/A	983	1947	174,586	220	1977	285,165	134	2007	255,187	33
1918	N/A	771	1948	176,364	203	1978	288,577	136	2008	258,892	23
1919	N/A	581	1949	182,638	152	1979	308,085	123			
1920	N/A	603	1950	180,955	164	1980	301,635	103			
1921	N/A	350	1951	185,244	175	1981	296,848	84			
1922	N/A	476	1952	186,463	209	1982	230,025	51			
1923	N/A	510	1953	188,692	161	1983	214,661	62			
1924	N/A	556	1954	177,425	139	1984	219,727	80			
1925	N/A	520	1955	184,239	157	1985	218,112	57			
1926	N/A	584	1956	200,807	172	1986	209,638	49			
1927	N/A	487	1957	219,151	152	1987	213,532	67			
1928	N/A	392	1958	269,076	167	1988	225,422	49			
1929	N/A	476	1959	288,560	173	1989	234,459	48			

Total number of metal/nonmetal mining fatalities from 1900 through 2008: **23,570.**

Note: Metal/nonmetal operations include mills (metal, nonmetal, and stone), sand and gravel, surface (metal, nonmetal, and stone), underground (metal, nonmetal, and stone). Sand and gravel miners are included starting in 1958. Office workers at mine sites are included starting in 1973.

From MSHA web site (10/7/09) at http://www.msha.gov/stats/charts/chartshome.htm

www.ingramcontent.com/pod-product-compliance
Lightning Source LLC
Chambersburg PA
CBHW081829170526
45167CB00007B/2756